# TECHNOLOGY, TRADITION, AND THE STATE IN AFRICA

Jug, now in the British Museum, bearing the arms of England and the badge of Richard II; discovered at a war shrine in Kumasi when the British defeated the Ashanti in 1896.

# TECHNOLOGY, TRADITION, AND THE STATE IN AFRICA

JACK GOODY

John R. Goody

*Published for the*
INTERNATIONAL AFRICAN INSTITUTE
*by*
OXFORD UNIVERSITY PRESS
LONDON IBADAN ACCRA
1971

*Oxford University Press, Ely House, London W. 1*

GLASGOW NEW YORK TORONTO MELBOURNE WELLINGTON
CAPE TOWN SALISBURY IBADAN NAIROBI DAR ES SALAAM LUSAKA ADDIS ABABA
BOMBAY CALCUTTA MADRAS KARACHI LAHORE DACCA
KUALA LUMPUR SINGAPORE HONG KONG TOKYO

SBN 19 724184 0



*Printed in Great Britain by*
*The Eastern Press Limited, London and Reading*

To
Esther

# CONTENTS

# PREFACE

It is the thesis of the present work that the nature of ' indigenous ' African social structure, especially in its political aspects, has been partly misunderstood because of a failure to appreciate certain basic technological differences between Africa and Eurasia. It is these differences that make the application of the European concept of ' feudalism ' inappropriate. But the problem is not only historical; in many areas ' traditional ' African social structure exists (in a somewhat modified form) precisely because the rural economy has not greatly changed. It is not only the comparative analysis of historians and sociologists that needs to take cognizance of these facts, but also the decisions of planners, developers, and politicians (both reforming and conserving).

In writing this monograph I have received help and advice from many colleagues: John Fage, Daryll Forde, Max Gluckman, Thomas Hodgkin, George Homans, Edward Miller, M. Postan; all have helped me to avoid some of the errors that an individual must make when he steps outside the narrow compass of his professional ' discipline '. John Fage asked me to discuss the subject of the first chapter at a seminar in London in May 1962; it was subsequently published as ' Feudalism in Africa? ' in the *Journal of African History*, iv (1963). Thomas Hodgkin invited me to read further papers on the military side at his seminar in Oxford. The second chapter was first presented to Professor Postan's seminar in economic history in Cambridge. Subsequently I rewrote it to give as the St. John's College Lecture at the University of East Anglia in 1968 and published it in *The Economic History Review*, xxii (1969). My thanks are due to the editors for permission to reprint these two papers in a revised form.

The drawings of the harness belonging to the Chief of Kpembe were made by Anna Craven when she was working as a research assistant with my wife, Esther Goody. I am also grateful to my wife for her help with the index and proofs.

St. John's College — Jack Goody
Cambridge

*November 1969*

# CHAPTER 1

# FEUDALISM IN AFRICA ?

WAS feudalism a purely Western phenomenon? Is it a universal stage in man's history, emphasizing replacement of kinship by ties of personal dependence which further social development required? If it is neither a universal prerequisite nor yet exclusively Western, what are the conditions under which it is found? A host of such questions are raised by the continual use, both by historians and sociologists, of the term 'feudal' as a description of the societies they are studying.[1] Here I want to inquire into the implications and value of the concept as applied to African social organization.

First used, apparently, in the sixteenth century,[2] the word feudal has since served an astounding variety of purposes, in everyday speech as well as in the writings of historians and sociologists. The primary referent is of course to a particular historical period, to Western Europe between the ninth and thirteenth centuries, to the social systems that on the one hand superseded the Roman Empire and the 'tribal' regimes which destroyed it, and that on the other hand preceded first mercantile and then industrial 'capitalism'. But the term has also been used of innumerable societies other than those of medieval Europe. A recent survey (Coulbourn, 1956) includes a comparison of feudalism in Japan, China, Ancient Egypt, India, the Byzantine Empire, and Russia. Nor is this simply an editorial quirk.[3] Many earlier writers on Japan had written of its feudal institutions[4]; Marcel Granet entitled his study *La Féodalité chinoise* (1952); Pirenne and Kees discuss the question in dealing with

[1] For an illuminating treatment of some of the general problems that lie behind this discussion, the reader should turn to Evans-Pritchard, 1961.

[2] In the sense of 'pertaining to the feudal system'. Of related words, some like feudary were used much earlier while others like feudalism were neologisms of the nineteenth century. The historian's discovery of the feudal system dates from the time of Cujas and Hotman in the sixteenth century. See Pocock, 1957: 70 ff.

[3] See also Stephenson, 1942: 1–2.

[4] See also Bloch, 1961: 446–70, and Boutruche, 1959: 217–97.

Egypt; Kovalevski and Baden-Powell do the same with regard to India and Vasiliev for Byzantium.

Historians are not the only persons to use this term in a comparative context. Social anthropologists have employed it in an equally all-embracing way. Roscoe and others have seen the Baganda as 'feudal', Rattray the Ashanti, Nadel the Nupe of northern Nigeria. Indeed it would be difficult to think of any state system, apart from those of Greece and Rome, upon which someone has not at some time pinned the label 'feudal'. And even these archaic societies have not been left entirely alone. Feudal relationships have been found in the Mycenean Greece revealed by the archaeologists and epigraphers, while it is generally agreed that one element in medieval feudalism was the institution of *precarium* of the later Roman Empire.[5]

Unless we assume the term has a purely chronological referent, then, or unless we are to take our smug refuge in the thought that persons, events, and institutions defy comparison because of their uniqueness, the use of any general concept like feudal, more particularly concepts like fief or client, must have comparative implications. Marc Bloch realized this when at the end of his classic study he wrote, 'Yet just as the matrilineal or agnatic clan or even certain types of economic enterprise are found in much the same forms in very different societies, it is by no means impossible that societies different from our own should have passed through a phase closely resembling that which has just been defined. If so, it is legitimate to call them feudal during that phase.' (Bloch, 1961: 446.)

There is then a measure of general agreement that 'feudal' should be used in comparative work. Here I want to look briefly at the ways in which it has been employed in the African context. For a large number of political systems of the 'state' type have been called 'feudal', and it seems pertinent to try and find out what the authors are getting at.

There are times when it seems as if people who work in the

[5] The *precarium* was a grant of land to be held by someone during the pleasure of the donor: the land was a boon (*beneficium*) granted as the result of the prayer (*preces*) of the recipient (Stephenson, 1942: 7; Pollock and Maitland (2nd ed.), 1898: i, 681 n.1). This practice has been the subject of an extensive discussion over the relative influence of German and Roman institutions upon feudal Europe.

non-European field use the term 'feudal' in the same spirit that led the composers of the *chansons de geste* to link the histories of their own petty kingdoms on the Atlantic seaboard with the great civilizations of the Mediterranean world; new-comers, upstarts, *nouveaux riches*, thus acquire the aura of respectability that tradition imparts. The danger in this is apparent. If the term has high status in the comparative study of society, there will be a tendency constantly to widen its range of meanings for reasons other than those of analytic utility. Moreover, an attachment to Western European models may turn out to be not the embrace of respectability but the kiss of death, just another version of the old pre-Copernican fallacy of the universe revolving around the earth.

One major difficulty in using the term for comparative purposes is that, even for historians of Europe, it has many meanings. In his introduction to the translation of Bloch, Postan writes of a recent Anglo-Soviet discussion on feudalism in which the two sides 'hardly touched at a single point. The English speaker dwelt learnedly and gracefully on military fiefs, while the Russian speaker discoursed on class domination and exploitation of peasants by landlords' (Bloch, 1961: xiii). These different views represent variants of two rather broader categories of approach which Strayer has summed up in the following words: 'One group of scholars uses the word to describe the technical arrangements by which vassals become dependants of lords, and landed property (with attached economic benefits) became organized as dependent tenures of fiefs. The other group of scholars uses feudalism as a general word which sums up the dominant forms of political and social organization during certain centuries of the Middle Ages' (1956: 15).

One can discern two trends in the narrower technical use of the term feudal. The first points to its derivation from 'fee', and hence to dependent land tenure.[6] The second emphasizes the lord-vassal relationship; it was to draw attention to this aspect of 'feudalism' that Pollock and Maitland suggested the term be replaced by 'feudo-vassalism' (1898: i, 67). In general, however, the core institution of feudal society is seen as vassalage associated with the granting of a landed benefit (fief), usually in return for

[6] e.g. Fustel de Coulanges, 1890: xii.

the performance of military duties.[7] In Max Weber's usage, feudalism is 'the situation where an administrative staff exists which is primarily supported by fiefs' (1947: 322).

The wider use of the term also has its variants, each with its own body of supporters. Apart from the loose popular turn of phrase that sees all types of hierarchical status (with the possible exception of slavery and bureaucratic office) as 'feudal', there are two main trends, one relating to political structure, the other to economic conditions. Political factors are stressed in the Coulbourn symposium, where Strayer summarizes this view in the following words: 'Feudalism is not merely the relationship between lord and man, nor the system of dependent land tenures, for either can exist in a non-feudal society. . . . It is only when rights of government (not mere political influence) are attached to lordship and fiefs that we can speak of fully developed feudalism in Western Europe' (1956: 16).

The thesis that feudalism is essentially a locally centred form of government is clearly connected with the existence of fiefs. Dependent tenures create (or recognize) a local administration of the fief-holder and those who inhabit his estate; they constitute a devolution of powers and are associated with a weakly centralized government that depends upon vassalage to provide military support.

The other line of thinking stresses the economic aspects and sees these as characteristic of a type of productive system. This was of course Marx's approach. He saw feudalism as one of the stages of pre-capitalistic economies, a 'natural economy' which preceded and led into the commodity market system. Changes in the division of labour were accompanied by different forms of property ('the stage reached in the division of labour . . . determines the relations of individuals to one another with respect to the materials, instruments and product of labour'): the first form is tribal property, the second the communal and state property of antiquity; the third form is feudal or estates property (1845–6: 115–19).

[7] Although fiefs are usually thought of as based upon the tenure of land, money-fiefs also played an important part in the West; they were of even greater significance in the Crusader Kingdom of Jerusalem (Runciman, 1960: 5). Kosminsky (like others before him) has pointed out that the bulk of manorial incomes took the form of money-rents rather than 'feudal' services. See Kosminsky, 1947, and the review by Postan 1950–1: 119–25.

Finally, the various political and economic features are clearly linked together in practice (though the analytic nature of the connexion is open to discussion), and there is a comprehensive approach that tries to define feudalism in terms of a number of these associated institutions. Of such a kind is the conclusion to Marc Bloch's study of feudal society, where he begins the section entitled ' A cross-section of comparative history ' with the words ' A subject peasantry; widespread use of the service tenement (i.e. the fief) instead of a salary, which was out of the question; the supremacy of a class of specialized warriors; ties of obedience and protection which bind man to man and, within the warrior class, assume the distinctive form called vassalage; fragmentation of authority . . . such then seem to be the fundamental features of European feudalism ' (1961: 446).

Each of these approaches to feudalism has been used by different authors in analysing the pre-colonial regimes of Africa. But the most explicit attempts to compare these political systems with medieval Europe have been in the work on northern Nigeria and the Interlacustrine Bantu, especially in the accounts given by Nadel of Nupe and by Maquet of Ruanda.

A section of Nadel's study, *A Black Byzantium* (1942), is actually entitled ' The Feudal State ', and here the author describes the manner in which tribute was collected and how the kingdom was divided into units of different sizes (' counties '), each comprising ' a town with its dependent villages and *tunga* which were administered as fiefs through feudal lords or *egba* ' (p. 117). These lords, who were recruited from the royal house, the office nobility, or the Court slaves, were eligible for promotion to more lucrative fiefs, although they continued to reside in the national capital. They constituted a ' feudal nobility ' who played an important part in raising military forces for the king, especially through their personal ' clients '.

The institution of clientship (*bara*-ship), which is widespread in the states of northern Nigeria, consists in a voluntary declaration of allegiance for the sake of political protection and often provides the basis for the formation of factions. There are a variety of forms, some of which involve military service, but Nadel sees the institution as essentially comparable to the *patrocinium* of Imperial Rome and medieval Europe (1942: 122–3).

In his recent study of one of the Hausa states, M. G. Smith analyses the changes that have occurred over the last hundred and sixty years in a rather similar political system. In the Hausa kingdoms, he writes, government 'is conducted through a system of ranked and titled offices known as *sarautu* . . . each of which can be regarded as an exclusive permanent unit, a corporation sole. These titled offices are characterized by such attributes as fiefs, clients, praise-songs, allocated farmlands, compounds and other possessions . . .' (1960: 6).

In Zaria, as in Nupe (but unlike most of the Hausa states), offices were not generally hereditary, except for kingship and the major vassal chiefships. The power and ambitions of fief-holders were controlled by the fact that they were clients of the king. Clientage (again there were a number of forms of *bara*-ship) is defined as 'an exclusive relation of mutual benefit which holds between two persons defined as socially and politically unequal and which stresses their solidarity' (1960: 8).

There is an interesting difference between the accounts of Nupe and Zaria. Whereas Nadel deliberately speaks of 'a feudal system', as Lombard does for the near-by Bariba of northern Dahomey (1957, 1960), Smith is content to use 'fief' and 'client' as analytic tools without making any overall comparison with medieval Europe.

We find another similar situation in recent accounts of the states of the Interlacustrine Bantu of East Africa. In his study of the Ruanda, *The Premise of Inequality in Ruanda* (1961),[8] Maquet describes the political system as a feudal structure. He defines a feudality as an organization 'based on an agreement between two individuals who unequally partake in the symbols of wealth and power culturally recognized in their society. The person who, in that respect, is inferior to the other, asks the other for his patronage, and, as a counterpart, offers his services. This is the essence of the feudal régime . . .' (1961a: 133). The term 'agreement' here is preferred to contract because the freedom not to enter into clientship was illusory, since no Ruanda could afford to live without a lord. Clients might belong to the ruling estate of cattle-keepers (the *Tutsi*) or to the subordinate group of

[8] See also his general discussion, 1961b.

agriculturalists (the *Hutu*)[9] but in both cases the transaction was established by the loan of cattle to the inferior partner, a transaction which Maquet regards as equivalent to the allocation of a landed fief in feudal Europe (1961a: 129, 133).[10] But he claims that the fief itself is only rarely found in Africa, because unlike medieval Europe, the tenure of land is not based upon Roman law (1961b: 294). By this I think the author means that a political superior (such as a king) does not 'own' the land in the same way as a feudal lord in Western Europe, i.e. in the same way a *Tutsi* 'owns' cattle, or the Nupe king 'owns' tribute. Hence the characteristic feudal formula of *Y* holding land of *X* (*tenere terram de X*) could not be applied. There is a certain truth in this observation, but it should be remembered that in England, at least, the formula of dependence seems to have been applied to a number of very varied conditions. The bundle of rights which we may think of as complete ownership of land was not always distributed among the actors and groups involved in precisely the same way and the nature of dependence differed in each case. Moreover, such relationships always contained an element of mutuality. From one point of view all higher contracts depended upon the performance of the basic agricultural tasks, and from the evidence concerning the inheritance of land at the village level it would seem that here the idea that conquest put all rights in the hands of the Norman conquerors was something of a fiction. Whatever the legal position on this abstract level, the medieval system in practice appears to display some similarities with African land tenure, especially in states like Nupe.[11]

Clientship in East Africa is also discussed by Lucy Mair in her recent book dealing with the political systems of that region (*Primitive Government*, 1962) and in a general paper on 'Clientship

[9] i.e. what Lombard speaks of as the lord-vassal (*liens de vassalité*) as well as the patron-client relationship (*liens de clientèle*) (1960: 11). More usually vassalage implies military service; according to Stephenson the term should be restricted to 'an honorable relationship between members of the warrior class' (1954: 250, n. 61); see also Boutruche, 1959: 293 ff.

[10] Mrs. Chilver points out that it bears a closer resemblance to the Early Irish form of cattle-clientage, *celsine* (1960: 390).

[11] For a discussion of the medieval situation, see Pollock and Maitland (1898: i, 234; ii, 4 ff.); for Africa, see Gluckman, 1945; for the concept of a bundle of rights, see Sir Henry Maine, *Ancient Law* (1861) and *Dissertations on Early Law and Custom* (London, 1883), 344.

in East Africa' (1961).[12] As a minimum definition she suggests: 'a relationship of dependence not based upon kinship, and formally entered into by an act of deliberate choice' (1961: 315).[13] She goes on to discuss the institution as it existed both among the cattle-keeping Ruanda and Ankole, and among the predominantly agricultural Ganda and Soga. Ruanda clientship was established by a cattle transaction, initiated by the would-be client with the words 'Give me milk; make me rich; be my father'; but the relationship was entered into because of a universal need for protection rather than for purely economic reasons. Among the Ganda and Soga on the other hand, society was divided into landlords and peasants, the former being subordinate territorial chiefs chosen by the king, who had control of unoccupied land. Mair speaks of the relationship established by the transfer of rights in land from chief to peasant as analogous to that of patron-client (a 'passive clientage', since personal service is absent); but she reserves the use of the term in the strict sense to the relationships between the king and his territorial rulers ('client-chiefs') and to that between an office-holder and his personal retainers ('private clientship'), from among whom client-chiefs were sometimes recruited (Mair, 1961: 322–3: Fallers, 1956: 135).

Mair also observes that clientship is a basis of social differentiation in two ways, firstly because 'it creates formally recognized relationships of superiority and subordination, defined by other criteria than seniority',[14] and secondly because in some societies such as Buganda 'it is the main channel of social mobility' (1961: 325).

We earlier saw that historical approaches to the definition of feudalism could be roughly classified into the technical and the general. These two sets of studies of societies in northern Nigeria and in the Interlacustrine region have been concerned with institutions similar or analogous to those which are the subject of the technical analysis of feudalism, namely clientship (or rather vassalage) and fiefs. But whereas Nadel and Maquet feel impelled to describe the societies they have studied as 'feudal',

[12] For two stimulating studies of particular societies with clientship institutions, see Southall, 1956 (for the Alur), and Fallers, 1956 (for the Soga).

[13] Residual categories often give rise to difficulties of analysis and this may prove to be so in the present case, where clientship is defined negatively in respect to kinship.

[14] See Fallers on the Soga (1956: 230).

Smith, Fallers, and Mair make at least as adequate an analysis without introducing the concept at all. This second approach seems preferable as a procedure. It is simpler; it minimizes the inevitable Western bias; and it helps to avoid the assumption that because we find vassalage (for example), we necessarily find the other institutions associated with it in medieval Europe. It is just these supposed interconnexions which comparative study has to test rather than assume.

I turn now to discuss the way in which the general approach to the study of feudal institutions has been used in work on Africa, beginning with the political. Strayer emphasizes that in medieval Europe many governmental functions were carried out at the local level, and Coulborn goes on to suggest that feudal systems are 'a mode of revival of a society whose polity has gone into extreme disintegration', i.e. after the break-up of a great empire (1956: 364; see also Hoyt, 1961). Here is an instance where the Western European starting-point heavily influences the outcome of the analysis. If we are to take as characteristic of feudalism the features that Strayer discusses at the beginning of the volume (i.e. clientship, fiefs, locally centred government), then it would seem that these institutions cannot only be associated with the revival of government after an earlier collapse. Indeed the African material points clearly to the fact that they may also occur as chiefless communities develop more centralized governments.

Southall touches upon this problem in his study of the Nilotic Alur of East Africa. His general purpose is to examine the 'process of domination' by which Alur dynasties become rulers of neighbouring, chiefless communities. In the course of his book, he discusses 'the embryonic political specialization of the Alur', which he sees as intermediary between chiefless societies on the one hand and state systems on the other (1956: 234). On the one hand he sees the Alur as introducing 'a new principle in the regulated allegiance of one man to another without any kinship bond existing between them' (p. 234), as in the relation of chiefs with their subjects, of nobles with their domestic serfs, and of chiefs with their various dependants. On the other hand, the Alur have no bureaucracy. 'The embryonic administrative staffs possessed by chiefs in their envoys and courtiers never acquired any formally defined administrative or jural powers, and remained occasional agents, advisers and confidants of their chief' (p. 240). Thus this type of political system, he concludes, is intermediary between what Fortes and Evans-Pritchard (1940) spoke of as state systems with an administrative organization (type A) and segmentary lineage systems (type B). Southall calls it the 'segmentary state' (as contrasted with the 'unitary state') and distinguishes as one of its most characteristic features the way in which both local and central authorities exercise very similar powers.[15]

In a section entitled 'The temporal and spatial range of the segmentary state', Southall goes on to consider other societies of this type, among which he includes the Ashanti, the Yoruba, Anglo-Saxon England, and 'feudal France in the eleventh century' (pp. 252–6).[16] While he does not specifically equate feudal and segmentary state systems, he does include eleventh-century France as an example of the segmentary state and his analysis of this kind of system bears directly upon a central problem of the Coulborn symposium on feudalism, namely, the question of political centralization.[17] And it is clear from the Alur case that we may expect to find such locally centred regimes developing with an increase in administrative centralization as well as after the disintegration of an even more centralized system. This evidence suggests that the latter hypothesis is one of the European-derived variety that we have to beware of. True, the sort of breakdown and build-up of central organization that occurred in Western Europe after the withdrawal of Roman rule is likely to give rise to specific institutional forms which warrant comparative treatment (for this situation is not in itself unique). But such comparison needs to include societies like the Alur

[15] Six characteristics are listed by Southall (pp. 248–9), but the distribution of power is basic to them all.

[16] As Stenton notes, the conquest state of Norman England did not display the same local independence in governmental matters that characterized feudal regimes elsewhere in Europe (1961: 5, 12–15). Unlike Maitland, Stenton emphasizes the differences in the English social system resulting from the Norman Conquest and asserts that 'only the most tentative of approaches had been made before the Conquest towards the great feudal principle of dependent tenure in return for definite service' (1961: 123). On the other hand, as Barlow remarks, 'most of the features associated with true feudalism can be found in the Old-English Kingdom' (1961: 11).

[17] Maquet has a rather different line-up. He sees the coercive sanctions of governments as operating permanently (i.e. in states) or intermittently (i.e. in non-states.) Feudality makes its appearance in both types of system, but not in states of a despotic kind.

which are in the initial process of centralization as well as those which are undergoing a secondary process. Only in this way can a satisfactory attempt be made to isolate those institutions that are linked with one process rather than the other, and those that are associated with both.

Clientship (as Southall, Maquet, and Mair see it) can certainly occur in both these situations. Indeed Southall and Mair regard this form of personal dependence as an essential element in the development of centralized institutions (Southall, 1956: 234; Mair, 1962: 107 ff.). Mair sees the elements necessary for this process as present even in a chiefless society like the Nuer, an extreme example of polities of type B. The factors crucial to the development of kingship (and so of government which revolves round a single centre-pin) are two: the belief that ritual powers are hereditary and the ability to attract and keep a following. A privileged descent group, privileged perhaps by virtue of first arrival, is able to expand by attracting attached and client groups; individuals are able to increase the range and extent of their authority by acquiring followers other than their own kin (p. 122). By these and similar processes, diffuse government gives way to minimum government and eventually to yet more centralized forms. Or, as Southall puts it, kin-based, chiefless societies develop into segmentary states and these to unitary states. Certain of the institutions thought of as characteristic of feudal societies, namely, clientship and locally centred government, emerge in the second phase of this process.[18]

The discussion which I have outlined in this rather summary way has a direct bearing not only on the question raised in the title to this chapter but also to some wider problems that face historians and other students of African societies. But to these points I will return after briefly considering what I have called, for want of a better label, the economic approach to feudalism; that is, the approach which defines feudal institutions primarily with reference to economic features, in particular the mode of land holding.

Any degree of political centralization entails specialized roles

[18] In a similar way L. A. Fallers, using Weber's typology of authority and Parsons's schema of pattern variables, describes the authority structure of the pre-colonial Soga as particularistic in contrast to the universalistic type of social relations that characterize bureaucratic structures; personal clientship is of course a particularistic tie of this kind (Fallers, 1956: 238 ff.).

and hence some withdrawal of man-power from primary production into administrative activity. To this extent at least all chiefs and lords 'live off the land', and usually have rights in land of a more far-reaching kind. But other more specific features are sometimes regarded as characteristically feudal—certain fiscal arrangements,[19] the seignorial mode of estate management, dependent tenure itself. Here I shall confine my remarks to the approach often adopted by orthodox Marxists, more particularly by I. I. Potekhin in his paper, 'On Feudalism of the Ashanti'.[20] Potekhin writes that 'Feudal land ownership constitutes the foundation of feudal relations'. Land belongs to a restricted circle of big landowners, while the peasant pays rent or performs services for the right to cultivate his land. In Ashanti, he finds 'the exclusive concentration of land in the hands of the ruling upper strata', together with the conditional land tenure and hierarchies of dependence 'typical of feudal society'. I have earlier commented that the idea of exclusive land tenure is hardly an accurate representation of the medieval evidence; it appears even less satisfactory as an interpretation of the Ashanti situation in the nineteenth century. To deny that 'ownership' of land is exclusively vested in one stratum is not of course to assert that Ashanti (or medieval Europe for that matter) was a 'classless' (i.e. unstratified) society. It is simply to state, firstly, that the concept of 'absolute ownership' of land (as distinct from other means of production) is probably applicable only to a society dichotomized into slaves and freemen.[21] Secondly, the splitting of the total quantum of rights in land between two individuals or groups does not necessarily imply a simple relation of dependency; delegation upward or devolution downward would give rise to much the same overall pattern. And lastly, it should be borne in mind that in pre-colonial conditions in Africa land was sometimes of little economic importance; for relatively low

[19] Mrs. Chilver (1960: 385) discusses these fiscal arrangements in examining East African states.

[20] Presented at the Twenty-fifth International Congress of Orientalists, Moscow, 1960. For another discussion, see Davidson 1961: 33 ff., where he maintains that writers on African states have often called slavery what was in fact a form of feudal vassalage (p. 38); he speaks of 'African feudalism' (as found, for example, in Ashanti) as 'tribal feudalism' (p. 46). For a contrary view, see Maquet, 1961b: 296–8, 307–10.

[21] On the important analytical difference between land and other property, see Pollock and Maitland, 1898: ii, 2, and Goody, 1962: 292 ff.

population densities (as compared, say, with Europe and Asia) meant that, in many regions, land was not a very scarce resource and hence its tenure could hardly provide the basis of differentiation for the 'class' system.

Before I conclude the discussion of the way in which the broader approaches to feudalism have been applied, let me recapitulate my comments upon the narrower technical approach. I remarked earlier that at this point in our inquiries I could see no great profit (and possibly some loss) in treating the presence of clientship or fiefs as constituting a feudality (as for example, in the work of Nadel and Maquet), as against analysing these institutions without inviting any overall identification of these societies with those of medieval Europe (as for example in the work of Smith and Fallers). There seems even less to be gained from adopting the view which sees African societies as feudalities on the basis of wider political or economic criteria. Firstly, there is the ever-present ambiguity of the term itself; and then again the primary referent is to a particular period in European history, and an author employing an analytic tool of this kind tends to focus the whole analysis around the Western situation. The difficulties are nowhere clearer than in the writings of those who see the development of human society in terms of the stages so widely used in the latter half of the nineteenth century. Writers who adhere to the orthodox Marxist doctrine, formulated a hundred years since, are particularly apt to fall back upon the idea of a universal progression from tribalism to slavery, feudalism, capitalism, and finally socialism, each such stage being characterized by a particular set of social institutions. Most historians assume a rather similar scheme, either explicitly or implicitly, and, like some sociologists, tend to speak of tribal or kinship societies in a way that suggests that they too easily overlook the strength of their own attachments to family and tribe.

Of course, certain general trends of development in political, legal, and economic institutions are rightly accepted by most students of society and the study of these trends has often gained much from the approach associated with the names of Marx and Engels.[22] What blocks advance, here as in other fields of comparative studies, is a rigid attachment to particular European-based

[22] e.g. in the writings of V. Gordon Childe. For a general account of the influence of Marx on the social sciences, see Bottomore and Rubel, 1956.

schema, whether this be derived from an explicit ideological commitment or from an inability to see beyond our own cultural tradition.

To suggest that there appears little to be gained by thinking of African societies in terms of the concept of 'feudalism' implies a rejection neither of comparative work that includes European society, nor yet of the contribution the European medievalists can make to the study of African institutions. The last point first. Although historians of the Middle Ages are dealing with their own cultural tradition, they are mostly writing about a very different set of customs from those they have absorbed with their mothers' milk. While their analysis is sometimes inhibited by the problem of origins, by a preoccupation with medieval institutions as the germ of contemporary ones, such writers are, on the other hand, forced to consider a broader range of human experience than historians who deal with more recent times. In this task they have been greatly helped by the great legal historians who contributed so much to the study of the medieval period, and the extent of whose contribution was due in some measure to the wide interests of comparative jurisprudence in the latter half of the nineteenth century. For a direct line runs from Fustel de Coulanges and Maine to Vinogradoff and Maitland, all of them men who had a considerable acquaintance with ethnological studies as well as with varied historical material. For Vinogradoff, 'comparative jurisprudence is one of the aspects of so-called sociology, being the study of social evolution in the special domain of law'; it draws its material impartially from ancient and modern, civilized and primitive communities (1911: 580). In this way comparative jurisprudence formed a link between the study of social institutions in primitive, archaic, and medieval societies. And indeed when Vinogradoff came to list the major figures involved, he included lawyer-anthropologists like McLennan, Bachofen and Morgan, the great Semitic scholar, Robertson Smith, the Romanist von Ihering, as well as other major figures in the history of anthropology, such as E. B. Tylor and J. G. Frazer. Frazer's work has of course influenced many writers on medieval subjects, particularly in the literary field, and while not all the results have been entirely happy, the author of *The Golden Bough* can claim some credit for the theme of Marc Bloch's *Les Rois thaumaturges* (1924). Again the whole body of work on village communities, in particular the analysis of early Anglo-Saxon and Celtic society undertaken by Frederic Seebohm (1883, 1895, 1902), stemmed from this same tradition.[23]

Thus medieval studies have been influenced in a variety of ways by comparative jurisprudence, which in its turn has had a direct link with social anthropology, or the aspect of it that some of us would prefer to call comparative sociology. The link continues today. Students of anthropology still study Maine, Fustel de Coulanges and Vinogradoff, as well as the writings of the major figures in sociology. I make this point in order to suggest that the work of some of the outstanding medieval scholars (and of classicists too) has already come into contact with comparative sociology in one form or another, so that one can only expect gains of a limited kind when the insights derived from their work are fed back into studies of African societies.

But, while the reverse is perhaps even more true, Africanists certainly have something to learn from the studies of medieval historians. Firstly, the work of Maitland, for example, is full of acute observations on topics like inheritance, marriage, descent, which provide valuable material for comparative analysis, quite apart from the question of whether 'feudal' institutions were present in Africa. Secondly, when dealing with centralized political systems, the anthropologist often acts as a special kind of sociological historian. He records the tales of old men; he may even administer questionnaires, as Maquet did in his Ruanda study. But if it is the indigenous system he is interested in, he cannot do what he does for marriage, household composition, and the like; he cannot go out and observe an independent state in action. It follows that if the anthropologist has to reconstruct

[23] 'Mr. Seebohm's *English Village Community* . . . revealed to us, for the first time, the inner life of mediaeval England.' (W. J. Ashley, 'The English Manor', introductory chapter to N. D. Fustel de Coulanges, *The Origin of Property in Land* (1st pub. 1889), trans. by Margaret Ashley, 1891, pp. xii–xiii.) There have of course been many criticisms of the 'tribal' school: Fustel de Coulanges' essay on property is one of these. Seebohm's work 'profoundly shocked the learned world of the day' (Stephenson, 1954: 241, n. 23) and both Vinogradoff's *Villainage in England* (1892) and Maitland's *Domesday Book and Beyond* (1897) were partly inspired by the desire to refute his thesis that English history 'begins with the serfdom of the masses' (Seebohm, 1883: ix). For a more specific comment on Seebohm's use of sources, see Timothy Lewis, 1956–7.

the past in this way, then he should know something of the elementary procedures of the ordinary graphohistorian.[24] The prospect of this course of indoctrination need not be too overwhelming. As much nonsense is talked about the techniques of the historian as about the methods of the sociologist—and with somewhat less reason.

Today the positive achievements of the writers in the field of historical jurisprudence seem to lie not so much in their grand picture of the development of social organization (although this matter was certainly of considerable importance in their work), but rather in the study of particular customs and concepts in a wide comparative setting and in the attempt to establish interrelationships between the institutions which they had isolated in this way. As instances of this work I would cite Maine on the relationship between ancestor-worship and inheritance (1883) and Vinogradoff on the connexions between types of agriculture and family composition (1920: i, 292).

If we are to take up and develop the tradition of comparative work, which has been so neglected in recent years by historians, sociologists and anthropologists alike, then the best strategy at this stage is to avoid the kind of overall comparisons that are invited by words like tribalism, feudalism, capitalism. These abstractions make for too crude a level of analysis. Social anthropologists are sometimes surprised at the sort of statements historians make about the social organization of African societies, just as traditional historians often raise their brows at the kind of remark sociologists make about the Reformation or about medieval Europe. We must avoid not only historical particularism, but also ill-considered generalities.

So far progress in the comparative study of centralized

[24] I use this term to make a distinction with the ethnohistorian. Ethnohistory usually refers to a study of the past which involves not only written records but also oral traditions, informant's versions of recent events as well as data of an archaeological and ethnological character. As there is a danger among traditional historians of assuming that a widely based study of this kind is in some sense inferior to a study resting upon documents alone it seems desirable (particularly in the context of African studies) to use a word that avoids the opposition ' history-ethnohistory ' and at the same time indicates the limitations of a method which ties itself exclusively to writing. Hence ' graphohistory '.

institutions in Africa has not been great;[25] the material is complex and compared with the study of lineage systems, for example, it has had little systematic attention. All the more need then for a considered approach.

How should this be tackled? We need first of all to concentrate upon the analysis of particular institutions, so that we can try to select the other factors with which they are associated. This means more than simply offering a definition of, say, 'clientship'. The process of constructing analytic concepts should involve spelling out the range of social behaviour implied and the alternative possibilities for human action.[26] If possible, it should also allow for 'measurement'; that is, for an assessment of gradients of differentiation and thus avoid commitment to a simple typology. And refinements of this kind are necessary, whether we are interested in comparing the differences and similarities of various social systems, or in discerning the sequences and explanation of social change. Until more work of this kind is done, the question 'Feudalism in Africa?' has little meaning, except for polemical purposes.

There is, however, a related problem of historical interest where recent sociological studies can help to clear away some of the cobwebs. This is the problem of state formation. African historians, even the 'ethnos', have been more or less exclusively concerned with centralized political systems, and for understandable reasons; in general, societies without rulers are societies without knowable history, and hence without historians. In Fage's *Atlas of African History*, for example, the diagrams are almost entirely of states. Plotted on a map, these units show a clustered but discontinuous distribution which invites questions about their point of origin and manner of diffusion. These are

[25] Apart from monographs of theoretical interest and the general works by Fortes and Evans-Pritchard (1940), Schapera (1956), and Mair (1962), some interesting studies have been made on a more particular level. There is Beattie's analysis of checks and balances (1959), Colson's discussion of bureaucracy (1958), Gluckman's work on rituals of rebellion (1954), Richards's papers on the role of royal relatives (1961) and the social mechanisms for the transfer of political rights (1960).

[26] Boutruche makes a start on this by undertaking the study of 'ties of subordination' in Frankish Europe and outside. He concludes his book with the following remarks upon the general question of feudalism: 'La féodalité est médiévale. Elle ne doit rien à l'Afrique, rien sans doute à l'Asie, le Japon excepté. Elle est fille de l'Occident' (1959: 297).

reasonable enough questions. But to answer them we need evidence, and of this there is little. If we have to make do with guesses, then these should be as well informed as possible. Recent studies of African states make it clear that while increased centralization in the political system almost always results from conquest, it is not only in this way that states arise. The Alur, for example, extend their domination when neighbouring peoples invite their chiefs to come and rule over them; we find, in effect, an upward delegation of authority rather than the assumption of power by a militarily dominant group.

Nor is diffusion, peaceful or violent, the only means. For if we modify the dichotomy between chiefless and state systems by introducing the idea of a gradient of centralization, as has been suggested by M. G. Smith (1956), Easton (1959), Mair (1962) and others, then the nucleus of state systems can be discerned even among the lineages, age-sets, cult-associations, and other basic groupings of acephalous societies.[27] The question of diffusion then assumes less importance in the total picture, for these nuclei need only the right conditions to develop into more centralized organizations.

A variety of factors suggest themselves here. In West Africa I have been impressed with the apparent ease with which small-scale, temporary polities of a centralized kind arose around (or in opposition to) the raiders for slaves and booty during the period immediately prior to the coming of the Europeans.[28] Then again, while the exchange of goods took place among and across peoples who lacked chiefs, long-distance trade was certainly facilitated by the presence of rulers, and did a good deal to encourage their growth.

[27] I do not mean to suggest, as others have recently done, that the dichotomy between 'acephalous' and 'state' systems is 'almost useless'. Even if one restricts the discussion to the apparatus of authority (in the Weberian sense), there are clearly great differences as one moves between societies at the Tallensi-Nuer end of the continuum, and those at the Ashanti-Nupe end. 'Acephalous' systems are not without holders of authority, but their jurisdiction is limited in terms of the numbers of persons involved and the activities covered; moreover, the methods of social control differ in emphasis and in substance from those employed in centralized societies.

[28] The histories of Samori, Babatu, and, on a much smaller scale, Bayuo of Ulu (Dagaba, northern Ghana) provide examples from the Voltaic area in the late nineteenth century. Southall describes the case of a man belonging to an acephalous society who entered the slave trade and posed as a chief (1956: 235–6, n.1).

These investigations suggest that any idea of the diffusion of kingship or chiefship from a single source, Egypt or elsewhere, should be treated with great reserve. Of course particular state systems have spread and undergone modifications in the process, by conquest and by other means. But before one can seriously entertain a hypothesis of diffusion based simply on the existence of supposed similarities, the criteria of comparison need to be carefully specified. 'Kingship' itself is much too vague. And to add the epithet 'divine' does little to help. We would expect any monarchy in Africa to be strongly linked to the religious system, whether it arose in response to local conditions or was created by some process of diffusion.

It is partly this multiplicity of modes of state formation which makes the formulation of the main lines of political development so difficult. That the history of man in the various parts of the world has been marked by a number of broadly similar developments in political institutions is a proposition that few would now wish to deny. In general this progression is seen as one from less to more complex forms of organization; the doctrine that the manners of simple peoples represent decadent remnants cast off by more advanced nations, the customs of those who have fallen from God's grace, no longer occupies the serious attention it did in 1871 when E. B. Tylor published his notable contribution to the study of cultural evolution, *Primitive Culture*.

All would agree now that, despite the hopes of some Utopian writers, there has been a general change from small-scale acephalous polities to large-scale centralized states. Beyond this there is little consensus. The contemporary world situation is in itself a denial of the assumption of many Europeans that there would everywhere be an inevitable progression towards parliamentary democracy.

Then again there is the fact that not all social developments, even in the field of technology, have always been in one direction. For there have been many cases where the useful arts have fallen into decay[29] and where political systems have adopted less centralized forms. In Asia, Leach sees the Kachin of Burma as oscillating in a sort of cyclical fashion between acephaly and monarchy (1954). And in East as in West Africa, slave-raiding

[29] See Rivers, 1912. A most striking instance of this process was the apparent disappearance of writing in Greece between 1100 and 800 B.C.

gave rise to a number of temporary, small-scale, centralized polities which later collapsed. While often, as among the Gonja of northern Ghana, we come across the case of a conquest state which has become more diffuse and locally centred in its system of government. Other states have lost control of borderlands, allowing their subjects to re-establish an acephalous structure.

But despite these qualifications, and despite the many and often justified criticisms of the application of evolutionary doctrine to social facts, only a real flat-earther would now regard the overall history of political systems as static, cyclical, regressive, indeed as anything other than a process of elaboration.

These questions concerning the origin and spread of state institutions and the rise and fall of different types of polity are ones upon which recent research offers some help. The extent to which the developments in Africa resembled those that occurred in Western Europe is certainly worth pursuing and could even shed some light on the major historical problems that engaged the attention of Marx and Weber. But in this, as in other comparative work, we must start with less worn counters, with more operational concepts. Otherwise the embrace of feudalism, far from leading to a hypergamous union of the desired respectability, will only end in an unhappy hypogamy.

## CHAPTER 2

# POLITY AND THE MEANS OF PRODUCTION

In the first chapter I discussed the application of the term 'feudal' to pre-colonial African states from the formal point of view and concluded that there appears little to be gained by thinking of African societies in terms of the concept of 'feudalism'.

But the usage, and indeed debate, continues apace; the number of feudal states in Africa has proliferated even in the last few years, the authors of the accounts explicitly or implicitly rejecting my caveat about the use of this term.[1] I should make it clear that these doubts are not about the possibility of finding broad resemblances between the states of medieval Europe and those of pre-colonial Africa: there are plenty of similarities in the structure of monarchical systems of government, wherever they are found.[2] My queries had to do with the utility of any analysis that rested on such vague and all-embracing concepts, that approached the situation in terms of wide categories rather than sets of particular variables. But there are more precise reasons why the overall comparison with medieval Europe seems inappropriate and this has, initially, less to do with government and politics than with economics and technology. In my opinion, most writers about African social systems, particularly when they are dealing with class and government, have failed to appreciate certain basic differences between the economies of Black Africa and of the Eurasian continent, and this failure has led to superficial comparisons not only in relation to 'feudality', but also in relation to

[1] See Loeb, 1962; Beattie, 1964: pp. 25–35; Cohen, 1966; Lombard, 1965. For an extreme position, see Gravel, 1965. Gravel's article is based on field work in 1960–1; in it he 'purports to describe certain *specific* aspects of life on a small "manor" in eastern Rwanda, and to show how remarkably similar they are to the same aspects of life on a baronial manor of medieval continental Europe' (p. 323), especially with regard to the absence of 'true markets', the 'economy of subsistence' and the 'self-sufficiency of the community'. For a view much closer to my own, see Steinhart, 1967. In general these discussions place very little emphasis upon the basic technology of medieval Europe, the use of mills, animal traction, etc.

[2] For a discussion of some of these similarities, see my introduction to *Succession to High Office*, 1966.

land tenure, property, inheritance, marriage, descent groups, and other important spheres of social action.[3]

Before I continue, I add a disclaimer. The identification of 'feudalism' in Africa has been associated with the left rather than the right. Palaeo-marxists accept a fixed progression, inherited from their nineteenth-century predecessors, from tribalism through feudalism to capitalism, though a greater element of flexibility is introduced with the recent publication of some of Marx's writings.[4] Others regard the discovery of the same processes in Africa that earlier occurred in Europe as crucial to the recognition of African history as a proper subject for academic discussion. Such an approach seems to me understandable, often correct, but in this case, misguided. We are here concerned with the utility of analytic concepts and whether 'feudalism' illumines more than it obscures. I suggest that we need to take a closer look at the means and organization of production in Africa and Europe instead of tacitly assuming identity in these important respects.

There are three interrelated aspects of the society I want to discuss, the system of exchange (that is, trade and markets), the system of production (especially the ownership of the means of production) and the military organization (and especially the ownership of the means of destruction).

My thesis is that while the pattern of trade showed little difference from Eurasia, and while the military organization displayed some similarities, at least in the savannah country of West Africa, productive relationships did differ in certain major respects. And secondly that these differences are relevant not only to the understanding of the past but to the present as well, and they need to be taken into active consideration when formulating development programmes.

First I want to stress that the difference between Africa and Eurasia does not lie in the presence or absence of markets and, in some spheres of activity at least, of a market economy. Much writing about non-European economics is based on the island communities in the Pacific—of the Trobriands, Tikopia or Rossell Island. These communities are atypical in that certain primary features of the economy arise from the fact that they are

[3] See Goody, 1969.

[4] See Marx, 1964, and the comment by M. I. Finley, 1968.

small, relatively isolated groups rather than just 'simple societies'. The concept of non-monetary economics is hardly applicable to pre-colonial Africa, with the possible exception of certain hunting groups of minimal importance. Africa was involved in vast networks of wide-ranging trade long before the Portuguese came on the scene. For East Africa we have a late first-century sailors' guide to the trade along the coast, the *Periplus of the Erythrean Sea.* Long before the Europeans arrived there were trade routes from Madagascar up to the East African coast, through the Red Sea and into the Mediterranean, along the Persian Gulf to India, South-east Asia and Indonesia. Possibly there was a direct route by which cinnamon was brought from the Spice Islands.[5] By the time the Portuguese reached the eastern shores of Africa, the Chinese had already been very active there; before the development of the gun-carrying sailing ship on the Atlantic seaboard, the maritime commerce of the Indian Ocean made Western Europe seem like an underdeveloped area.[6] Indeed, the trade between Ethiopia, the Mediterranean and the Indian Ocean had much to do with the developments in the Arabian peninsula, possibly including the rise of Muhammad.[7]

In West Africa the medieval empires of the Niger bend were built up on the trade which brought salt, cloth, and beads south from the Sahara across to West Africa and took gold and ivory and slaves back to the Barbary coast and from there into medieval Europe. When the British defeated the Ashanti in 1896 they found a war shrine in Kumasi consisting of a bronze ewer and a jug. The jug now stands in the medieval section of the British Museum as one of the finest examples of English craftsmanship at that time. It bears the arms of England and the badge of Richard II, and is inscribed with the following proverbs:

He that will not spare when he may
he shall not spend when he would.
Deem the best in every doubt
till the truth be tried out.

[5] See Miller, 1969.

[6] Cipolla, 1965; Serjeant, 1963.

[7] The early trade in the Indian Ocean had important consequences for the population of Madagascar, the spread of Asian food crops to Africa, and for the changes in the social organization of the coastal areas of Eastern Africa and parts of the Arabian peninsula prior to the rise of Muhammad. Some aspects of these latter changes are discussed by Wolf, 1951.

If it were known, the story of how these vessels reached the tropical forest of West Africa would encapsulate much of the economic history of trans-Saharan trade.[8]

From the point of view of mercantile economy, parts of Africa were not dissimilar to Western Europe of the same period. Metal coinage was in use on the East African coast. In the West, currencies consisted of gold, brass, salt, but more especially cowrie shells which, coming as they did from the Maldive Islands off the south of Ceylon, filled most of the necessary attributes of money. In certain respects this was a monetary economy. Trade was highly organized and in kingdoms such as Dahomey and Ashanti important sectors of the economy were under state control, whereas in the savannah regions exchange was left largely in private (Muslim) hands. Most of the kinds of economic operations that were found in pre-industrial Europe were also to be found in Africa; even in the stateless societies of the interior, barter had been superseded by more complex forms of exchange, and production was rarely limited to subsistence alone; the extensive use of cowries from the Maldives and carnelians from Gujerat shows that they were all in some degree part of the economic system of the Old World. The impact of long-distance trade on social organization of course depends upon the degree to which productive activity is diverted to serve the purposes of external demand. My point is that, except in the special fields of the wine and wool trade, the differences between the external exchanges of Africa and early medieval Europe appear to have been relatively slight.[9]

If mercantile activity was not vastly different from that of medieval Europe, what about other aspects of the productive system? As we have seen, some writers have claimed that land tenure in African states was feudal in kind; others dispute this contention, denying the utility of the concept of a landed fief in Africa. Most of this discussion, which I reviewed in Chapter 1, has taken place on a politico-legal level. But there is one crucial and obvious difference which has been largely overlooked. It is a difference which means that African land tenure (and hence

[8] *Guide to Medieval Antiquities*, 1924, fig. 156. See also the fourteenth, century bronze ewer, in the same museum, which was ' the great war fetish of the Ashanti Nation ' (*Brit. Mus. Quarterly*, viii, 1933, p. 52). For the European side of the trans-Saharan slave trade, see Verlinden, 1955.

[9] For medieval trade, see Carus-Wilson, 1954.

vassalage and landed fiefs) was unlike that which obtained in much of Europe and indeed in much of the Eurasian continent generally; and it has to do with the means of production rather than with productive relations, though its influence upon these relations is of considerable importance. Basically Africa is a land of extensive agriculture.[10] The population is small, the land is plentiful and the soils are relatively poor. Moreover, one fundamental invention that spread throughout the Eurasian continent never reached Africa south of the Sahara, with the exception of Ethiopia. I am referring to that Bronze Age invention, the plough.

What effect does the plough have? In the first place it increases the area of land a man can cultivate and hence makes possible a substantial rise in productivity, at least in open country.[11] This in turn means a greater surplus for the maintenance of specialist crafts, for the growth of differences in wealth and in styles of life, for developments in urban, that is, non-agricultural, life.[12] In the second place, it stimulates the move to fixed holdings and away from shifting agriculture. Thirdly (and not independently) it increases the value (and decreases the availability) of arable land.

In Africa, then, there was little use of machines, even elementary ones; agriculture has meant hoe farming which was carried out

[10] The difference between extensive and intensive modes of agricultural production is clearly relative, and one that has to be related to the nature of the soils, the labour force and the terrain; shifting cultivation continued in less fertile and less accessible parts of medieval Europe long after the plough dominated the agricultural scene, and the same is of course true of Ethiopia today.

[11] In forests the plough clearly has many limitations. Nor does it improve vegeculture to the same extent as it does the cultivation of cereals. The main point however was effectively made by V. G. Childe. See also McNeill, in *The Rise of the West*: 'The harnessing of animal power for the labor of tillage was a step of obvious significance. Human resources were substantially increased thereby, since for the first time men tapped a source of mechanical energy greater than that which their own muscles could supply. The use of animal power also established a much more integral relation between stockbreeding and agriculture. Mixed farming, uniting animal husbandry with crop cultivation, was to become the distinguishing characteristic of agriculture in western Eurasia. It made possible a higher standard of living or of leisure than was attainable by peoples relying mainly or entirely upon the strength of merely human muscles' (1963: 25–6).

[12] Urban centres of course existed in pre-colonial Africa; they ranged from the agro-cities of the Yoruba to the polyethnic trading and administrative towns of the Saharan fringes and the coastal regions.

by men or women or both, depending upon the particular society. Indeed animal power, that drew the Eurasian plough, was not used for any other form of traction. One immediate reason was that the wheel,[13] though it crossed the Sahara, both in the West (as evidenced in the two-wheeled chariots liberally engraved upon Saharan boulders) and in the East (in Ethiopia and in the early Sudan), never penetrated pre-colonial Africa (or rather was never adopted there). Nor was this because of the lack of a metal technology. While Black Africa escaped the civilizing influence of the Copper and Bronze ages, the smelting of iron diffused rapidly from the Mediterranean down both sides of the continent.

In the East the technique of iron-smelting travelled to the Sudan, where Meroë has been described, with some exaggeration, as the ' Birmingham of Africa '; the metal began to be known there in the sixth century B.C., roughly the same time as iron was found in Ethiopia. From there it spread to Chad in the first century A.D., possibly brought across by horsemen using a long lance. In the West, iron was transmitted from Carthage and the Barbary Coast to the Niger towns in the third century B.C. The technique of iron-working, which followed later, spread through West and East Africa. By ' the first few centuries A.D. ' it had been introduced to Zambia by a number of small related groups of immigrants who brought not only metallurgy, but also food production and pot-making from the area west of Lake Tanganyika. Copper and bronze were employed in many parts of pre-colonial Africa, but before the coming of iron the extent of this use was negligible.[14]

The absence of the wheel meant that man was not only unable to make use of animal power, but of the power of wind and water as well. This is why the recent introduction of the lorry, the bicycle and the engine-driven mill has had such a revolutionary

[13] Animal disease was another factor limiting the use of the plough. Of Ethiopia, which may have obtained the plough from South Arabia or Egypt even in pre-Semitic times (about 1000 B.C.), F. J. Simoons has written: ' Where there are animals suitable for ploughing, both Cushites and Semites use the plough; but where, as along the Sudan border, these animals are excluded by disease, even Semites turn to the hoe or digging stick for preparing their fields ' (1965: 11). Iron-working however appears to have arrived from South Arabia at about the same time as writing, that is, in the fifth century B.C. (Anfray 1968: 352).

[14] See Mauny, 1952 and 1961.

effect upon the rural economy of Africa. But the lack of the wheel had another consequence for agriculture, since it limited the possibilities of water control. In the drier regions of the Eurasian continent the wheel has played a dominant part in raising water from wells to irrigate the land. Simple irrigation there is in Africa, as almost everywhere agriculture is practised. Some of the inhabitants of the settlement of Birifu (LoWiili) in northern Ghana channel the water from a permanent spring to run among their fields, and thus get two crops a year in place of one. The Sonjo of Tanzania practise more developed water control. Rice growing in the Western Sudan, and it should be remembered that some of the rice used here (*Oryza glaberima*) was domesticated independently of Asian rice in the Senegal-Mali region, demands yet more positive measures.

There are other means of water control that do not involve the wheel, that is, using various techniques of temporary storage. Methods of this kind did of course exist. Everywhere there was some improvement of natural pools. In Gonja and in neighbouring areas of northern Ghana there are many ancient cisterns hollowed out of the laterite; in the famous market town of Salaga, the city of 1,000 wells, these are cylindrical in form and do not seem wholly dependent upon surface water. But these storage systems are very different from the village tank of south-east Asia; while there is no lack of water in Africa, the problem of its distribution is enormous. And in terms of agriculture what is lacking, apart from the *shaduf* of the Saharan fringe, which uses the lever principle, is any mechanical device for drawing water, such as is used in the Middle East and even in the Saharan oases.

One further highly important effect of the technological gap between Africa and Eurasia lay in the military field. When the Portuguese spearheaded European expansion into other continents, they succeeded largely because of their use of gun-bearing sailing ships.[15] Through these they could dominate their African opponents who were armed only with sword, spear, and bow. But, by the end of the fifteenth century, when the expansion of

[15] I use the term 'gun' in a general way. At first they depended upon the cannon on their floating castles; later upon the hand-gun. See Cippola's useful discussion where he quotes Pannikar as saying that by 1498 'the armament of the Portuguese ships was something totally unexpected and new in the Indian (and China) seas and gave an immediate advantage to the Portuguese' (1965: 107).

Europe began, their guns were also far in advance of Asia as well. These technological innovations soon spread from Europe, just as simpler forms of gunpowder and ' cannon ' had earlier diffused there from China and the Middle East. But the way in which they did so is of great interest. Beachy notes that ' the Africans never seem to have learnt to make fire-arms as good as those of the Europeans, unlike the sixteenth- and seventeenth-century Japanese and Sinhalese, who soon achieved virtual parity with the Portuguese in this respect '.[16] Already in the mid-sixteenth century the Japanese were producing matchlocks and they soon were followed by the Koreans; Ceylon had become a centre of production by the end of the sixteenth century, when muskets and cannon were made at many different points in the Indian subcontinent (Cipolla 1965: 127–8). By the seventeenth century the inhabitants of the Malabar coast were exporting muskets to Arabia.[17] The reason for the failure of Africans successfully to take up the manufacture of this powerful new weapon is a simple one. They did not possess the requisite level of craft skill in iron-working. As a result, Africans were at an enormous disadvantage when the scramble for their continent began, since they had to fight against the very people who were supplying them with arms.[18]

[16] Review of B. Davidson, *J. African Hist.*, iii (1962), p. 510.

[17] See Cipolla, 1965: 127–8 and Al-Djamuzī's history in Serjeant, 1963: 117: ' These Malabaris are Muslims: the Munaibārī muskets are called after them '.

[18] The position was not quite as desperate as at first appears, since one European power was quite willing to benefit at the expense of another; the sources never entirely dried up. And even within the same political unit (e.g. the British-administered Gold Coast), the interests of merchants and administrators were often at odds. It should be added that, though sulphur had to be imported, there was certainly some manufacture of local gunpowder (e.g. in northern Ghana) but this was recognized to be of inferior quality (because it was not ' corned '). Equally, guns could be repaired. But there is no evidence of any extensive manufacture of components. Even today imported bicycle frame tubes are often used to replace barrels. The one exception may have been the late nineteenth-century Mandingo warrior Samory. In 1898 Nebout was reported to have been shown a small-arms factory ' capable of turning out three repeaters a week '. The gun is described ' as a wonderfully close copy of the French article, but not bearing near inspection as to its details, more particularly in the matter of rifling (Dir. Mil. Intell. to C.O., 14/1/1898 C.O. Confid. Print, African (West) No. 549). Legassick quotes a similar report (1966: 104). Peroz and Binger, who separately visited Samory in 1897, make no mention of locally-made guns; Henderson, who was captured in 1897, writes only of a cartridge factory. Local *numu* blacksmiths may have

What does all this add up to in socio-economic terms? Firstly, in Africa rights in land were less highly individualized than in most of Europe, partly because land was not a scarce commodity.[19] Among the Bemba of Zambia rights in the productive use of land, other than the small proportion under cultivation at any one time, hardly existed; the same is still true among the Gonja in northern Ghana. Under such conditions neither individuals nor kin groups bother to lay specific claims to large tracts of territory, since land is virtually a free good. Elsewhere, among the LoWiili and Tallensi of northern Ghana, for example, the population densities are around 100–200 per sq.m. and rights in land more highly developed; these rights tend to get tied down to small kin groups, such as minimal lineage segments, although residual rights are vested in larger descent groups which often see themselves as property-holding corporations. But in fact rights to these assets are divided up among the smaller units, which are themselves constantly splitting and reuniting, depending upon the distribution of the births of male children.

But, overall, people were thin on the ground. Even today, the total population for Black Africa is not much more than that of the United States, although the surface area is four times

learnt new techniques from the invaders, but it is difficult to see how the low-temperature forge could be used for the manufacture of guns from scratch, as distinct from the kind of repair work that Binger describes (1892: i, 191–2). In Europe the temperatures needed to emerge from the wrought-iron phase were first produced in the fourteenth century by using water-powered bellows for blast purposes, and this in turn depended upon rotary motion. But whether or not Samory succeeded in manufacturing firearms, his continued attempts to secure imported guns by exchanging captives or by more direct methods indicate that his efforts in this direction met with little success. Indeed right to the end the military economy of his empire-building was based upon the necessity of acquiring sufficient income from booty to import guns, powder and percussion caps (Goody, 1965: 75–8).

[19] The tendency for rights in land to be more highly individualized the greater the population density has been noted by Meek, 1946: 149–50. See also Jones, 1949: 313, and Goody, 1956: 37. Of course land shortage is always relative to the available resources of labour, etc. But the tractor can create a shortage more easily than the plough, the plough more readily than the hoe. In my view there is no doubt that the 'communal' nature of African land tenure must break down rapidly with the introduction of plough or tractor, just as the economy of larger households must tend to fragment when individual members acquire their own pay-packets, by selling their labour outside the domestic group.

as great. Not only was land plentiful, it was also less productive than in Eurasia, partly because of the technological limitations, and partly because tropical soils are often of poor quality. Moreover, since processes of soil regeneration (either by manure or by special cropping) were limited in nature, the fertility of land soon fell off. Under these conditions, the answer usually lay in moving one's farm (though not necessarily one's residence) to a new site, that is, in shifting cultivation.

The social consequences were two-fold. Politically, chiefship tended to be over people rather than over land; these a leader had to try to attract as well as restrain. The conditions for the forms of domination that obtained in the European Middle Ages hardly existed, except for slavery itself.[20] In slavery, labour is controlled by political force; in serfdom, economic controls, such as land tenure, are of equal importance. It is highly significant that only in Ethiopia, which had the plough, was there any landlordism in Africa; here in true medieval fashion, estates in land supported a nobility that filled the important offices of state, both in the staff and line organization, a nobility that was at the same time a leisure class in Veblen's sense. Besides the nobility one also found ecclesiastical landlordism—functionaries whose time was devoted to the glory of God (though individual commitment to the monastic life was often temporary rather than permanent in character) and who

[20] Edward Miller points out that there is another situation which leads either to serfdom or to plantation slavery. Both institutions can emerge under conditions where 'land is relatively plentiful and where it is necessary to prevent the escape of rent producers or labour producers which a chief or landlord requires. Perhaps an extreme form of this is the slavery found in pioneer settlements in certain parts of early medieval Europe.' In Africa labour requirements led to slavery but not serfdom; trading towns like Kano and Bida in northern Nigeria, or Salaga and Bole in northern Ghana, were surrounded by villages of slaves which supplied the ruling and commercial groups. Domestic slaves, dependent kinsfolk and clients filled other servile roles, but the supply of land and the degree of control made it difficult to exploit labour by anything other than slavery. In Middle America, despite ample land, the Spanish conquerors imposed on the population a system of peonage, which was something less than slavery. But the conditions of productivity (the plough, soil, crops) and control (guns and horses) were very different.

derived their 'living' from the church with which God had been endowed.[21]

If you have landlords, you can also have tenants and serfs; unfree tenancies mean little unless land is highly valued and your peasantry has nowhere else to go.[22] Under conditions of shifting cultivation, it means little. Slavery was important throughout most of Africa: war captives were given household or agricultural work to perform for their captors or their purchasers. But ties of subordination arose not out of shortage of land but as the result of purchase or conquest, thus giving rise to slavery rather than to serfdom.

Though there were no landlords, there were of course lords of the land—the local chiefs of centralized states, who, from the standpoint of food production, were in a sense carried by the rest of the population; we may either look at this as a return for services rendered or as the exploitation of the weak, for there is, I think, no real test. On the one hand chiefs could not be expected to sit around hearing complaints on an empty stomach. Their families did better than those of commoners; they often rode where others walked. But in general, due to the limited nature of the technology, to the relatively low differentiation in the terms of levels of consumption, it seems that standards of living, as measured by the usual tallies, were not markedly different. Gluckman has remarked of Zulu chiefship that one man can only eat a limited amount of porridge; the rest that he accumulates

[21] In Ethiopia the Church was the largest landowner after the Emperor. Some of these lands were worked by the monks; others were farmed by peasants. The landlord's rent, lay or ecclesiastical, could be enormous. At the end of the eighteenth century, Bruce noted that the tenants of Tigre usually surrendered at least half their crop, the landlord supplying the seed. But it was 'a very indulgent master that does not take another quarter for the risk he has run' (see Pankhurst, 1961: 193–5). A similar situation existed in Egypt with regard to the *waqf* lands. See for example Ibn Khaldun's reference to chieftains who 'are wont to build Mosque schools, shrines and almshouses and to endow them with Waqf [Mortmain] land . . .' (Issawi, 1950: 144). Of course monasticism could also be supported by other means than land, by 'taxes' on trade, or by providing services for travellers. But this was not, I think, possible to any great extent under pre-colonial conditions in Africa.

[22] The reduction of population in relation to resources can have the opposite effect on serfdom. When in the fourteenth century land in Europe became relatively plentiful (either through the Black Death or other causes), rents were halved in a decade, wages doubled and the institution of serfdom was greatly weakened.

has to be distributed.[23] Probably the maximum differentiation in terms of what Weber called ' styles of life ' was to be found in the trading states of the coast, in the empires of the Niger bend and in the emirates of northern Nigeria, where there were certain modes of behaviour that distinguished the urbanites of the capital (the dynasty, the merchants, the learned men, the specialist workers, and the hangers-on) from the rural population. But in most cases such differentiation was confined to chiefs themselves rather than spreading to the whole dynastic descent group, being a function of roles rather than social strata; and the most noticeable aspect of the difference lay in control over women and slaves, and guns and horses, rather than goods and land.[24] The exceptions were to be found in the coastal strips, where ruling and merchant groups financed themselves out of the European and Asian trade: there are certainly some sumptuary distinctions reported by travellers from Dahomey and Ashanti. But these differences do not approach those of medieval Europe, nor even of Ethiopia, where a series of sumptuary rules confined to the nobility the playing of certain musical instruments, the brewing of honey wine, and similar forms of behaviour that were by definition of high prestige.[25]

[23] Gluckman suggests that the failure of the southeast Bantu to develop more extensive political units before the time of Dingiswayo at the beginning of the nineteenth century may have been due to the limited technology and to the availability of land, so that there was possibly little point in building up power. ' The tribal economy was simple and undifferentiated; even in a good year the available technology did not allow a man to produce much beyond his own needs. There was little trade and luxury, so even a conqueror could not make himself more comfortable than he had been before. One cannot build a palace with grass and mud, and if the only foods are grains, milk and meat, one cannot live much above the standard of ordinary men ' (Gluckman, 1960: 157–68). In West Africa the economy was more developed in these respects, particularly the military economy in so far as it depended upon cavalry and the importation of firearms and gunpowder. But the general point still holds.

[24] In Hausaland, as elsewhere in the western Sudan, trading too produced great differences in wealth.

[25] Levine notes that the upper stratum of Abyssinian society ' created and shared what might be called a " gentry sub-culture ", even though it was not so differentiated from the general culture as was the case in Europe and China ' (1965: 156). It should be added that sumptuary laws in the strict sense often arise when certain individuals or groups are challenging the prestige behaviour of old-established classes; they are, Edward Miller points out, ' a defensive mechanism to help maintain the existing social hierarchy '. See for example Edward II's Ordinance against extravagant housekeeping, 6 Aug. 1316 (Stubbs, 1882: 238–9).

I have been suggesting that while there were local chiefships (a line organization) supported partly out of agriculture, partly from trade, there was nothing equivalent to estates in land of the European kind. I should perhaps modify this statement and say 'save only in some places and under limited conditions'. Because if you turn from line to staff organization, Colson suggests that the large-scale development of central government, which as Weber had argued involves the creation of appointive (as distinct from hereditary) office, was possible only in a few African states because of the land situation that I have described.[26] The southern Bantu failed to develop in this way. But some states, she maintains, succeeded. Among those that did were the highly centralized kingdoms of Buganda,[27] of Barotseland (Zambia) and of Dahomey. In each of these cases special conditions existed. In Buganda, the banana provided the basis for continuous cultivation; in Barotseland, there were the fertile lands of the Zambezi flood plain; in Dahomey there was the development of some plantation agriculture.[28] In each of these kingdoms you got a system of 'office estates' (rather than fiefs proper), that is, estates attached to the staff rather than the line organization.

Limited estates in land, sufficient to support appointive offices, were only rarely built up under conditions of this kind and in any case did not give rise to the kind of landlord-tenant (or serf) relationship characteristic of Europe. But was it possible to use benefices of a different kind in a similar way? In Ruanda, estates in cattle were employed to create relationships of subordination that resembled the cattle clientage of early Ireland.[29] Such estates are less easy to assign for the support of appointees to offices. They easily become fused with the personal property of the incumbent; and in any case, support by livestock is the formula for a very much looser polity than the predominance of appointed office suggests; it is difficult to centralize cows.

A more likely form of benefice to substitute for the estate in land is one based upon the allocation of income from tax or trade, which also provided one of the bases of support in western Europe. Few states of Africa were not involved in long-distance exchange.

[26] Colson (1958) speaks of the variable as the 'attitude to land' but the context of her remarks allows us to give this a more concrete reference.

[27] See Wrigley, 1957.

[28] And more importantly perhaps, the European trade.

[29] Maquet, 1961a: 129 ff.; E. M. Chilver, 1960: 390.

In the case of coastal trade with Europeans (as in Dahomey and Ashanti), goods could be channelled through the machinery of government. Trading privileges could be granted directly by the king, and taxation held few administrative problems.[30] But even in the savannah kingdoms where state import-export systems did not exist, the trading communities could be forced to contribute to the support of office-holders in a variety of ways —by market taxes, customs dues, and transport charges.

Other more problematic methods of endowment did exist. These lay in the field of military organization, which is the third aspect of the social system to be considered. Here, at least in the savannah country of West Africa, the military technology resembled that of feudal Europe, since war was dominated by the use of horsed cavalry. In this connexion it is relevant to turn to the thesis elaborated by Heinrich Brunner concerning the relationship of feudalism and cavalry in Europe. Brunner maintained that the great change from infantry to cavalry fighting came in the eighth century, some time after the Frankish king, Charles Martel, faced the Saracens and their horsemen near Poitiers in 733. In this battle Charles was victorious but lacking cavalry he was unable to follow up his success and Brunner held that the change to cavalry occurred shortly after this event and marked the beginning of feudalism. White (1962: 3) agrees that this 'was essentially military, a type of social organization designed to produce and support cavalry'. But he doubts whether it was the Saracen invasions themselves that produced the change, for the redistribution of lands had already begun by 732 and moreover the Muslims made little use of horses. It was rather the advent from the East of a technological device that made mounted warfare possible, namely, the foot-stirrup. The traditional Frankish arm had been infantry. The stirrup however enabled the horseman to gain greater control on horseback so that he could wield a sword or charge his enemy with lance at rest. With the coming of the stirrup, the *francisca*, the distinctly Frankish battle-axe, and the *ango* or barbed javelin, disappear. The long sword comes into use and so too does the spear with a heavy stock and a spur below the blade to prevent too deep a penetration. This kind of horsed warfare led in turn to

[30] Busia, 1951: 79 ff.; Wilks, 1966: 215–32; Herskovits, 1938.

developments in the sphere of defence, especially the use of increasingly heavy armour.

This new form of combat required a considerable capital outlay; the military equipment for one warrior cost about twenty oxen, or the plough-teams of at least ten peasant families (White, 1962: 29). To maintain cavalry of this kind there had to be some pay-off for those making the investment. In France the redistribution of church lands appears to have met this need, providing an income, a pasture, and a training ground for 'the feudal class' which, as White remarks, 'existed to be armed horsemen, cavaliers fighting in a particular manner which was made possible by the stirrup'.[31] The original and basic knight's service was mounted shock combat; it was around the cavalry that notions of chivalry arose, that there developed a knightly ethic based upon *loiautee* to leader and *proesce* in arms. 'The feudal aristocrat might, indeed, be a ruler, but this was incidental to his being a warrior' (1962: 30).

The horse penetrated early into Africa up the Nile valley, into the Sudan, and across the Sahara to the region of the Niger and Lake Chad. So later did the stirrup—the toe-stirrup in Ethiopia, but elsewhere the foot-stirrup. But there was no later changeover to heavy cavalry equipment; the heavy lance never replaced the javelin, and the coats of kapok, mail and Koranic leather charms worn by the horsemen were mainly a protection against infantry rather than against opposing cavalry.

The failure to develop heavy cavalry was no doubt partly due to the weaker economic base, which inhibited the accumulation of the required capital. But in any case lighter cavalry was characteristic of the Berbers of North Africa, who also were not directly faced with Christian arms (White, 1962: 36); and furthermore the lighter cavalry of the Turks and other Muslim powers had the advantage of speed, in attack, in pursuit, and in escape.[32] The added mobility placed a premium on raiding rather than warfare. In Europe the cavalry reimbursed themselves in part from loot, slaving, and from the ransom for their more valuable captives; at other times they had their estates to sustain them. In

[31] White, 1962: 28. Parts of White's thesis seem too rigidly tied to the stirrup, but the connection of European feudalism with cavalry is clear. For a criticism of White's thesis, see Hilton and Sawyer, 1963.

[32] Cipolla, 1965: 91.

Africa, the cavalry were maintained partly by taxes on trade, by tribute and protection payments, but largely from booty. The booty did not primarily consist of material goods, for there was little to seize from the average farmer except his family. The pay-off was in human booty, captives to be sold as slaves. Booty was indeed part of the productive system of the ruling class. A measure of this close interdependence of cavalry and raiding is the situation that obtains today in the eastern Gonja capital of Kpembe, the twin city of Salaga. A cavalry headquarters at the end of the last century, it now boasts only one horse; yet on the wall of every entrance-hut hangs the dusty and disintegrating harness that tells of former glories.[33] The establishment of colonial over-rule spelt the end of the external raiding and the internal control, thus undermining the economic position of the ruling estate. When the British authorities seized power and monopolized the use of force, the Gonja had no further use for the animal on which their domination had largely depended. The economy and power of the ruling estate largely collapsed; their authority was restricted by the new rulers. But they retained enough to carve out a place for themselves in the colonial and independent regimes that followed.

Although there was no development of heavy cavalry in Africa, the military system had nevertheless some basic similarities with that of medieval Europe. Here I refer not so much to the trappings of war, some of which were indirect borrowings, but rather to associated features of the political system. For the very nature of cavalry warfare imposes certain patterns upon the social organization.

In the first place the cavalry usually provide the ruling strata, since they also provide the means of destruction, the means of conquest, the means of booty. Unlike military technologies based upon the bow and arrow or upon iron infantry weapons, there is a built-in stratification between horse-soldier and foot-soldier.[34]

[33] See Braimah and Goody, 1967.

[34] The horsemen are particularly useful for fighting peasants and collecting captives, but the archers still had an advantage at close range, behind defensive positions. The layout of buildings and settlements in the Western Sudan was profoundly influenced by cavalry. (For examples of fortifications, see Binger, 1892: i, 93, 171.) With the advent of the horse, the cavalry became the offensive or predatory arm of states, whereas the bow was employed mainly for defence, though from the later sixteenth century it was supplemented by imported firearms, initially in very short supply.

Secondly, horsed warfare demands a considerable investment of skill as well as of money—a long training in horsemanship which again tends to make for differences in 'class consciousness', for ideas of nobility and the knightly ethic. While these were less marked with light cavalry, they were not altogether absent.[35] Thirdly, the investment of skill and capital was not a free gift to the nation. Not only was there a pay-off in booty but there was also a political increment. Since military power was centred upon the cavalry, they were well set to demand a share in political power. This they might achieve by acting as kingmakers, or else as strong divisional chiefs. But in West Africa the diffuse government characteristic of European feudalism often took the different form of a system of succession by which the paramountcy passed between the various segments of the ruling estate.[36]

Cavalry states have certain common features both in feudal Europe and in the circum-Saharan Africa, though the difference between heavy and light cavalry remains important. But most of Africa was unsuitable for the horse which is more sensitive to tsetse fly than other livestock. Not only is the animal absent from all forest areas; it never penetrated south of the Sudd marshes of the Sudan. Consequently the structures of East African kingdoms display a number of differences from those found in the West. Even in Buganda, where war was an economic activity, it required little investment of skill and goods;[37] no large-scale endowment of land and peasant services was needed to support a permanent body of knights. Nor was there the sharp dichotomy between horse-owners and agriculturists that existed in the savannahs of West Africa, though the rigid stratification that marked Ruanda and some other interlacustrine states of central Africa had a similar root in the division between a ruling estate whose economy was based on livestock, and a conquered group who worked the land.

I have argued that the social systems of Africa and Eurasia displayed some broad differences which were closely linked to

[35] However, the exclusiveness of classes was limited by the open marriage system and by the limitations of the agricultural technology; for a discussion of this point, see Goody, 1970.

[36] See Coulbourn (ed.), 1956. For an analysis of rotational systems, see Goody, 1966b.

[37] War brought 'women, slaves, cattle and ivory into the elaborate system of gift-exchange centred on the Court' (Chilver, 1960: 386–7).

the productive technologies, more especially in the rural sector. In Africa, there was virtually no use of rotary motion nor of animal traction; there was almost no application of machinery to cultivation, to water control, or to transport. Practically the only machine was the loom.

# CHAPTER 3

# POLITY AND THE MEANS OF DESTRUCTION

In the last chapter I began by discussing differences in the means of production in Africa and Eurasia and ended by pointing out certain similarities related to the means of destruction—especially some of the implications of a cavalry system for medieval Europe and for the Western Sudan.

In this chapter I want to examine differences in the political systems within West Africa and my aim will be to link such differences with their technological base. I would stress that I am not trying to establish a one-way causal chain between two variables. But I am trying to explore the implications of certain broad differences in technology between Africa and Eurasia on the one hand and within Africa on the other. Included in these differences in technology is the question of access to European goods, such as the handgun. It is unnecessary to remind the reader that, in talking of pre-colonial states, I do not imply that such states were uninfluenced by the advent of Europeans to the coastal areas of Africa. The import of guns and the external demand for slaves encouraged the war-like proclivities of centralized governments and consequently the nature of their interaction with other peoples.

In pre-colonial times, over the long period when European activity was largely confined to the coast, there were to be found in West Africa most of the broad varieties of political organization that comparative sociologists have tried to distinguish in Africa, with the exception of the hunting hordes.[1] Acephalous societies and complex states existed side by side in the same field of social relationships. Yet by and large the productive systems of the whole area were very similar. Traders it is true found greater security in the states and most entrepôts were situated within their boundaries. But trading activities extended well beyond these limits.

[1] Though hunting continues, there are no hunting 'tribes' in West Africa; the nearest are the pygmies of the Congo forest and some hunters on the fringes of the empty quarter of the Sahara.

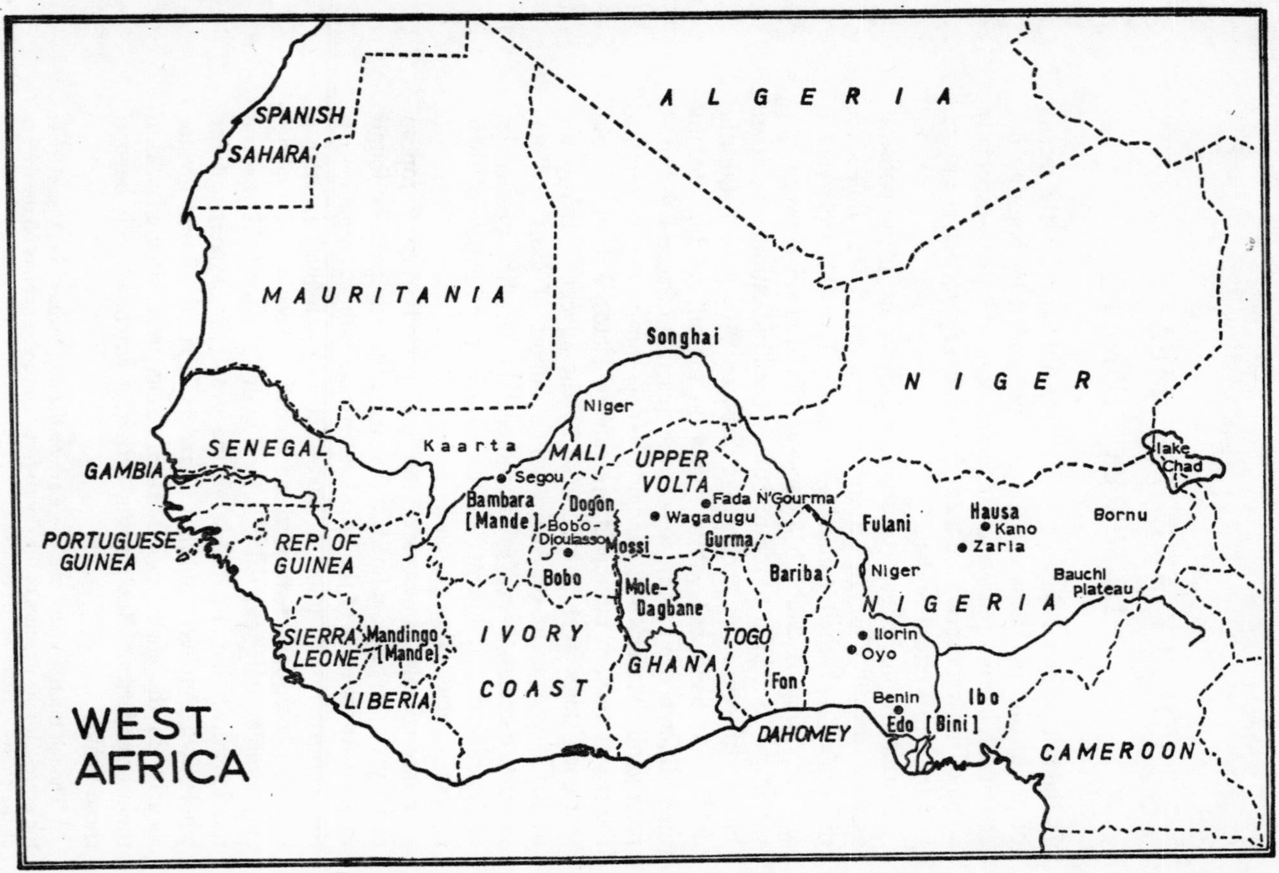

Map 1. West Africa

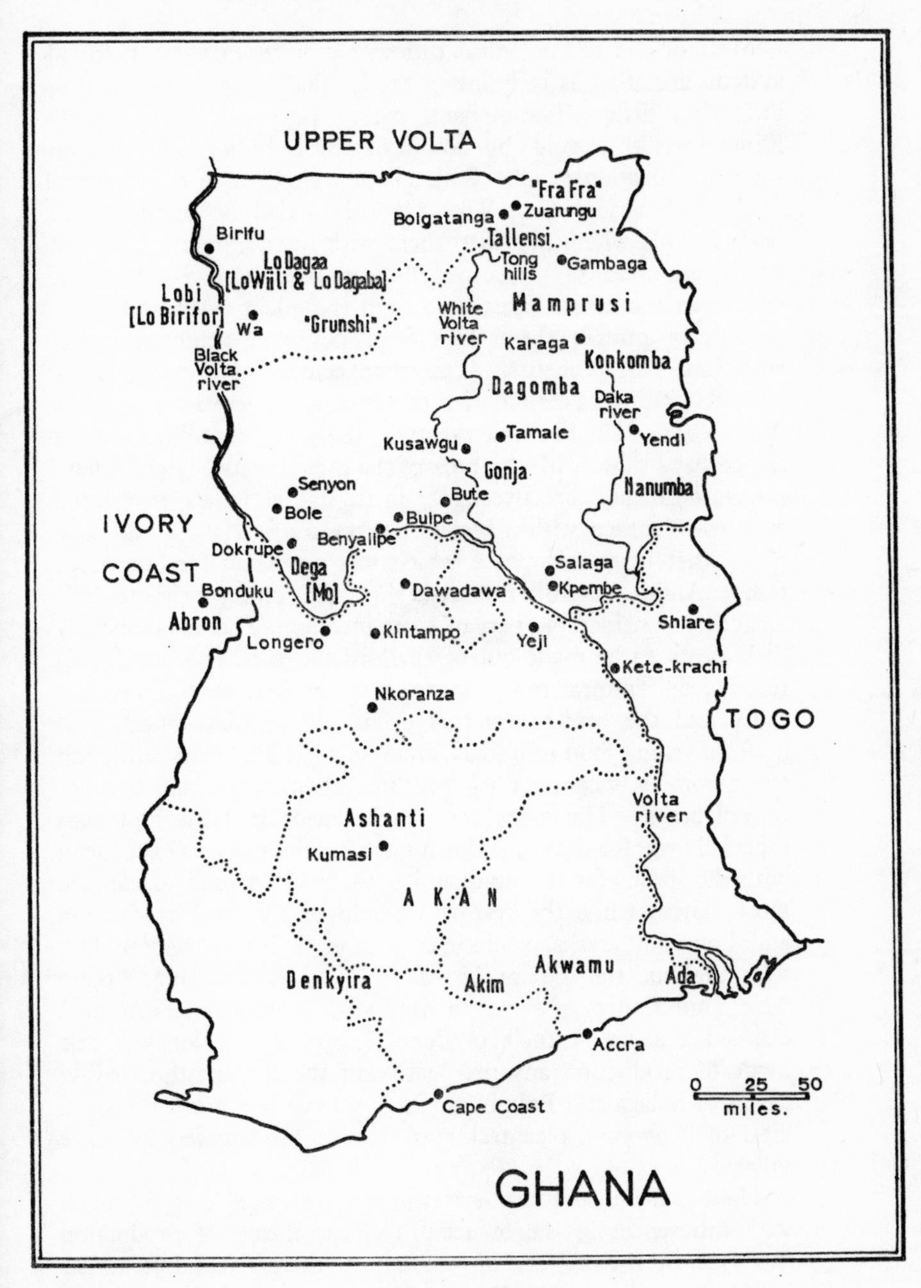
UPPER VOLTA
"Fra Fra"
Zuarungu
Bolgatanga
Tallensi
Tong hills
Gambaga
Birifu
Lo Dagaa
[Lo Wiili & Lo Dagaba]
Lobi
[Lo Birifor]
Wa
"Grunshi"
White Volta river
Mamprusi
Karaga
Konkomba
Black Volta river
Dagomba
Daka river
Tamale
Yendi
Kusawgu
Gonja
Nanumba
Senyon
Bute
Bole
Buipe
IVORY
COAST
Benyailpe
Dokrupe
Dega
[Mo]
Salaga
Kpembe
Bonduku
Dawadawa
Shiare
Abron
Longero
Kintampo
Yeji
Kete-krachi
Nkoranza
TOGO
Volta river
Ashanti
Kumasi
AKAN
Akwamu
Denkyira
Akim
Ada
Accra
0 25 50
miles.
Cape Coast
GHANA

Map 2. Ghana

Most of the technological differences between these political systems are of a relatively minor kind. The modes of production differ but little. The Ashanti nation vigorously exploited its mineral wealth in gold; but so did nearby Lobi, a people with no state organization at all. Both peoples engaged in trade, metal work, and agriculture. The Ashanti carried on an extensive trade in kola which provided them with many imports from the north; but such commerce was also practised by non-centralized groups in the Ivory Coast who used the silent trade and other devices to protect themselves from powerful neighbours and ambitious entrepreneurs. The main centres of exchange were located within the boundaries of states and there is, I think, a direct connection between the two; the more centralized states are perhaps those with the bulk of the export trade. Even where the state did not directly engage in trading activities, it ensured an area of peace within which market transactions could take place, that long-term peace which was necessary for long-term trade. And it was from this trade that such states (or their chiefs) derived a considerable part of their income. There was usually little profit to be made out of landlordism or indeed out of any form of agricultural production, given the low level of productivity and the limited internal exchange facilities—apart, that is, from selling food to traders, from kola (which grows wild) and some from early commercial crops in Dahomey and later in other coastal areas. The gains were to be made in trade and war, especially war for slaves; of this human booty some was exchanged with Europeans for the guns and powder to be used to capture more slaves, while the rest was employed for food production and household tasks, sometimes in the political 'household'. 'Throughout the Fulani kingdoms', writes Stenning, 'there were communities of slave cultivators . . . whose Fulani masters claimed a share of their produce' (1959: 17). Nor was this mode of production any prerogative of the Fulani; the trading towns of Salaga and Bole in northern Ghana had much the same kind of economy, a central market town surrounded by slave villages.

These various political systems are correlated not so much with differences in the ownership of the means of production (nor yet in the objects of production themselves) but rather in the ownership of the means of destruction and in the nature of those means. In one sense, a superior military technology was the productive system of the ruling strata, since it led to the acquisition of slaves, other booty, and taxes on trade. But productivity in the military field clearly differed from productivity in the agricultural and industrial sphere, since such activity necessarily resulted in the impoverishment of others.

In West Africa we are basically concerned with four military technologies: (i) bow and arrow; (ii) spear and sword; (iii) horse; (iv) gun.

The bow and arrow, the characteristic weapon of the acephalous peoples of the area, was used not only in warfare but also in hunting. It was even more closely identified with an individual than his main productive instrument, the hoe. After death it enshrined many aspects of his social personality.[2] As an instrument of production the bow of course persisted even after the introduction of hoe agriculture and livestock; as a weapon it persisted even after the introduction of other weapons or means of delivery. The Dahomean army had its contingents of archers; so too did most of the savannah states.[3] While it certainly lost a great deal of its importance in later technologies, in European history it made a temporary comeback in the shape of the English long bow during the Hundred Years War.[4]

The bow and arrow is essentially a democratic weapon; every man knows how to construct one; the materials are readily available, the techniques uncomplicated, the missiles easy to replace (though more difficult with the introduction of iron that

[2] See Goody, 1962: 83, 202, for an account of its social and symbolic importance among the LoDagaa; the bow, the quiver and the arrows represented the dead man and his possessions in the various funeral ceremonies.

[3] The bow is part of the regalia of some Gonja divisions (Kpembe and Bole; about the rest, my information on this point is uncertain) as well as of the Bambara states of Segou and Kaarta (Monteil, 1924); in a parallel way the gun plays a highly significant part in Dahomean rituals of succession (and earlier, a hoe handle). I should add that in Gonja the spear, with its cavalry associations (there was very little shield and spear warfare among the infantry in West Africa), was a yet more important part of the regalia.

[4] The tactics of the English in the Hundred Years War were modified by the lessons learnt in the Welsh (1277–95) and Scottish wars (1296–1328) of Edward I and Edward II. 'Taught by the events of Falkirk and Bannockburn, they abandoned the old idea that battles were won solely by the charge of armed horsemen. Success, it had been found, depended far more upon the judicious use of archery' (Oman, 1924: ii, 111). It was this change, partly brought about by the guerrilla tactics of the British resisters to English colonial conquest, that lay behind the English victory at Crêcy (1346).

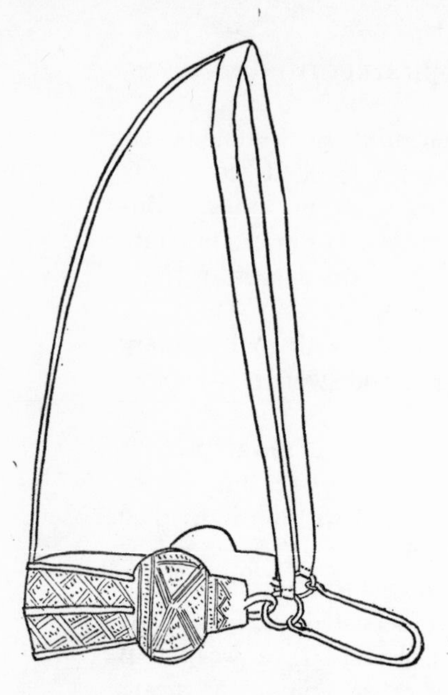

Nose band—standing martingale attached to chin piece. Brass, with tape straps.

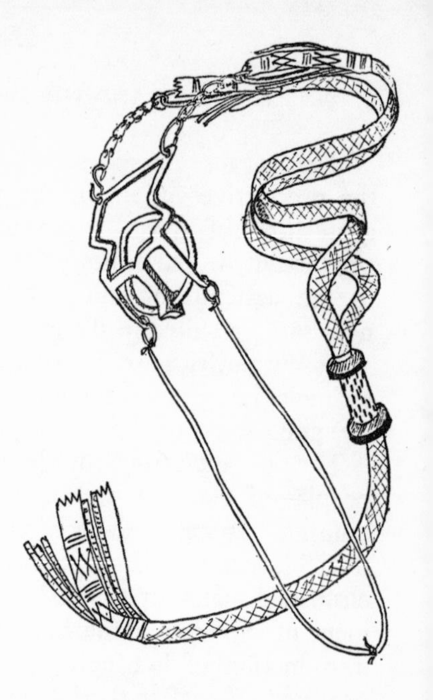

Bit and reins. Ring = curb (chin piece). Reins leather.

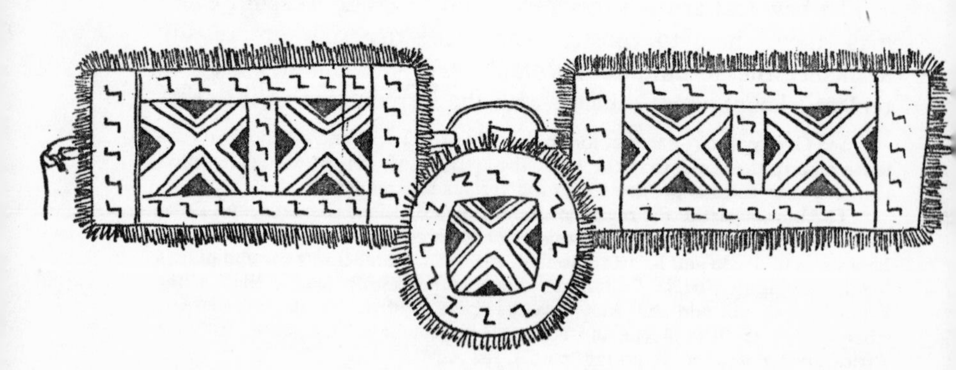

Chest piece, attached at either end to saddle. Turquoise cloth with embroidery as side pieces.

Harness in possession of the Chief of Kpembe, Gonja, Northern Ghana, in 1964; drawn by Anna Craven.

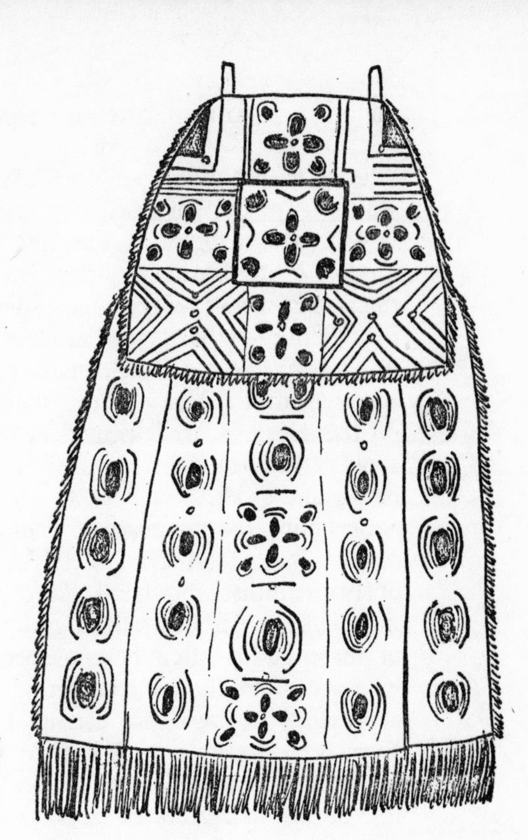

Side piece hung from saddle—one of two, covering withers. Turquoise cloth with red, blue, green, gold embroidery.

Head piece. Black designs on turquoise leather, with brown, yellow and black leather strands; blue, purple and black wool fringes. Has cheek strap underneath.

affected even hunting peoples like the Hadza of Tanzania and the Bushmen of the Kalahari).

With technologies of the bow and stone-tipped arrow, any kind of centralization is almost impossible. But with the introduction of metals, kingdoms are on the cards. First, because the distribution of the raw materials is uneven and involves systems of exchange (often long-distance trade) which can be brought under control. Many of the states of the savannah zone lived off the dues that trade provided, while at the same time offering some services to the trader in return, especially the maintenance of law and order.

Secondly, the processes of manufacture are relatively complicated. In some areas of West Africa we find special kin-guilds of blacksmiths who hand down their traditions among their members; and in centralized groups such as Mossi the members of these guilds often have a special relationship with royalty, who are often their major patrons. But elsewhere (among the acephalous LoDagaa, for example) smiths are not so restricted. Even here, however, such individuals have a special role to play in the maintenance of peace, perhaps to counter-balance their role as manufacturers of arms, though arms are instruments of production as well as destruction, killing animals (which one eats) as well as men (whom one does not—except in a ' ritual ' fashion).

But quite apart from the increase in productivity which the use of metals offers (and hence the possibility of maintaining a more complex administrative system), it is possible to supervise the technology itself, the weapons and the trade in weapons; some effective central control of force becomes feasible for the first time.

That it does not become inevitable is obvious; one factor militating against central control of weaponry is the very widespread distribution in Africa of a low-grade iron ore known as laterite. Other metals are not so widely spread, but then other metals did not have this kind of military importance since Africa south of the Sahara had no real Bronze or Copper age.

As we have noted, iron first spread from North Africa around 500 B.C. and evidence of the vigour with which it was later developed can be seen in the massive slag heaps of Meroë in the Sudan. On the western side, the techniques of iron-working crossed the

Sahara at approximately the same time. Iron provided blades for the hoe, points for the arrow and tips for the spear. Productivity was raised; so too was the potential for domination. State systems appear to have proliferated. But even so both productivity and domination were of a limited kind. Productivity, because of an extensive agriculture and the absence of basic technological inventions such as the plough and the wheel. Domination, because the infantry that was the basis of war was specialized neither in its formations (there were no professional soldiers, though the Zulu later approached such a system), nor in its weaponry, which was also the weaponry of the chase. Some iron throwing-knives were used but the sword was never of great importance (except ritually); the basic infantry weapon was the spear or javelin, with the shield for defence.

Iron weapons became of much greater importance when they were combined with a method of delivery that harnessed the power of the horse. In the ancient Middle East the original military use of horsepower was in drawing chariots; this gave mobility to the infantry, transporting warriors to and from the scene of battle. Such vehicles, while they penetrated the Sahara, hardly crossed it. But the horse did, at least as far as the marshy Sudd region of the Sudan and throughout the savannahs of West Africa; elsewhere the tsetse appears to have halted further expansion.

The cavalry warfare of the Sudanese zone bore some distinct resemblances to the warfare of medieval England; different observers have made the comparison of Europe with Bornu (Urvoy, 1949) and with the Fulani states (Davidson, 1961: 35) where, as in the Eastern Sudan, Islamic heraldry and horsemanship doubtless had a direct influence upon the Muslims of this region. But despite the wide diffusion of the foot-stirrup, the cavalry did not employ the heavy lance and sword of European armies; it had light armour, mail or quilted, and it used the stabbing spear and lighter swords; often the horse was just a means of transport to war rather than in war, though it was also used in raids to charge down the infantry of acephalous peoples.

But in one organizational feature at least the military demands upon the social system were similar both in Sudanese Africa and in medieval Europe; these were certain demands of the horse itself. This animal could not be employed in agriculture (for there was neither plough nor wheel) and was maintained largely for military purposes; it also served as a means of personal transport, but never for goods, a role reserved for the humbler and tougher donkey, sometimes for the bullock, and, in the desert, for the camel. The horse was the noble animal in every sense and one reserved for the nobility and its hangers-on. It was reserved for the nobility, not because of special sumptuary laws, but because of the large investment in the means of destruction that horses entailed. Chivalry, the horse culture, had a politico-military base.[5]

Under certain conditions, it is possible to build up a centralized cavalry, a military arm directly under the control of the monarch.[6] It is also possible, in pastoral societies, to find the ownership of horses distributed throughout society. But in West Africa, as in medieval Europe and most other parts of the globe, horses were the possessions of a politically dominant estate that was usually of immigrant origin and had established its domination over a

[5] It had of course a yet closer connexion in Europe, a connexion which dwindled from war and transport to horse-racing, fox-hunting, show-jumping, activities which are still identified with the upper strata. It was Jonathan Swift who satirized the horse culture of the English aristocracy in his account of the Yahoo in Gulliver's travels among the *Houyhnhnms*: 'when I asserted that the *Yahoos* were the only governing Animals in my Country, which my Master said was altogether past his Conception, he desired to know, whether we had *Houyhnhnms* among us, and what was their Employment: I told him, we had great Numbers; that in Summer they grazed in the Fields, and in Winter were kept in Houses, with Hay and Oats, where *Yahoo* Servants were employed to rub their Skins smooth, comb their Manes, pick their Feet, serve them with Food, and make their Beds. I understand you well, said my Master; it is now very plain from all you have spoken that whatever Share of Reason the *Yahoos* pretend to, the *Houyhnhnms* are your Masters' (*Gulliver's Travels*, New York: Random House, 1934, p. 234).

[6] A dependent cavalry does occur. The Mamelukes of Egypt, who governed the country from 1250 until the Turkish conquest of 1517, formed such a group. The last of the Ayyubite dynasty, Malik al-Salīh, Najm al-dīn Ayyub, was noted for his purchases of slaves (often Turkish) to reinforce the army; these formed a permanent soldiery, for whom barracks were built and whose enfranchised officers, members of the court, soon took over the government and ruled for the next two and a half centuries. As Ayalon (1956: x) has pointed out, it was their continued commitment to cavalry and their failure to adapt to firearms that led to their defeat by the Turkish janissaries. The latter had already adopted muskets from Western Europe; in 1589 they revolted and took over at least partial control of the Turkish empire from the Ottoman dynasty. The effects on the system of succession to high office were dramatic (Goody, 1966: 21).

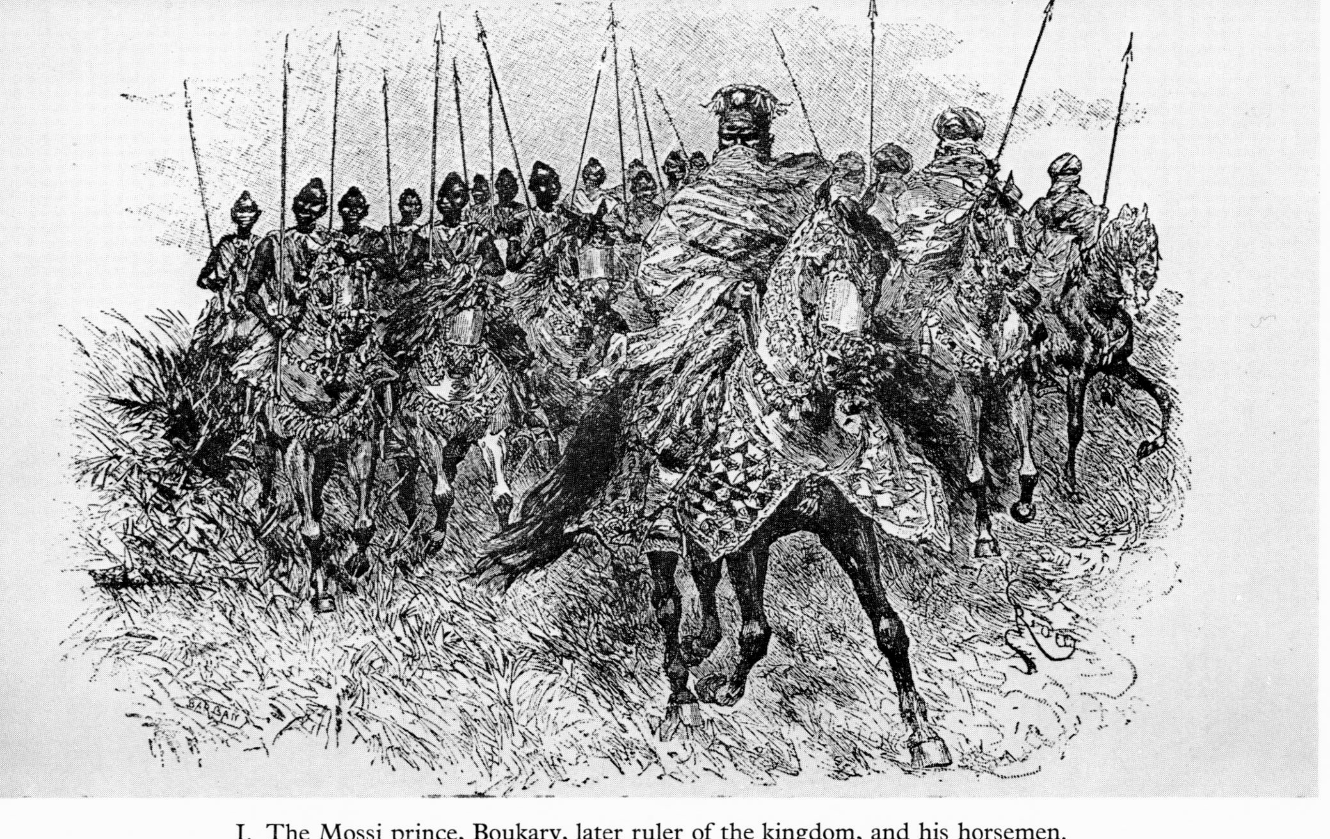

I. The Mossi prince, Boukary, later ruler of the kingdom, and his horsemen.

The French explorer, L. G. Binger, met Boukary in 1888, when he was attempting to reach the capital at Wagadougou. At this time Boukary was living on the frontiers of the kingdom, raiding the non-centralised peoples for slaves (L. G. Binger, *Du Niger au Golfe de Guinée* Paris, 1892, vol. i, p. 455).

II. The horsemen return, bringing their captives.

In his second stay with Boukary, Binger describes how the chief secretly despatched two raiding parties at night. One returned with seventeen slaves, the other with five, together with a donkey carrying salt and a little cotton cloth. Three

land of peasant farmers.[7] Being part of the system of domination, being essential to the maintenance of the position of the monarchy, the ruling estate were in a position to make demands as well as give support; the one entailed the other. Because of the nature of the productive system, the demands could be sumptuary only to a limited extent and were for power rather than styles of life. The ruling estate held many of the large territorial offices in the land and provided a constant check to the power of the monarch. In some places, Northern Nigeria for example, elaborate governmental institutions were developed, but there was no real autocracy.

The military capability of centralized states is related to certain general features of their social organization. A number of writers have commented upon the differences between the forest and savannah states of West Africa, Lombard (1957 and see also 1965) draws a specific comparison between the Bariba kingdoms and the well-known state of Dahomey to the south. The Bariba he describes as 'feudal', referring to the decentralized nature of its political organization, its emphasis on clientship and (though this is not a feature of European feudalism) to its circulating system of succession. Dahomey was highly centralized, with its national army, its complex census machinery, its formidable administration. When Richard Burton visited the country in 1863, he commented forcibly upon the contrast between the soldier-like discipline of Dahomean despotism (1864: i, 116), and the less autocratic ways of the Gold Coast and Gabon. 'Here [the peasants] never dream of such ownership [of tables etc.]. . . . More to make them feel his power than to ameliorate their condition, [the king] will not allow them to cultivate around Whydah coffee and sugar cane, rice and tobacco. . . . A caboceer may not alter his house, wear European shoes, employ a spittoon-holder, carry an umbrella without leave, spread over his bed a counterpane, which comfort is confined to princes, mount a

[7] The words aristocracy and nobility imply a small group of rulers and one that was initially defined by 'noble' birth. These implications are rarely justified in Africa. Babatu of northern Ghana was a horse-trader who got side-tracked into free-booting for slaves (Krause 1928). But this piratical free-lance activity could have led to the formation of a cavalry-based state, the ruling group of which would subsequently have been defined by birth. Here it is the noble animal, the *Houyhnhnm*, that enobles the *Yahoo*; doubtless a number of states originated in this way, rather than as the result of the fission of an existing dynasty.

hammock, or use a chair in his own home; and if he sits at meat with a white, he must not touch knife or fork . . .' (1864: i, 180–1). The power and authority of the king was clearly much greater than in the savannah kingdoms.

Much the same contrast is observable between the Ashanti of southern Ghana and the Gonja (and other states) in the north of the country. While the federal institutions of the Ashanti have been emphasized by many writers, it is clear that power was increasingly centralized in the capital of Kumasi, under the command of the Ashantihene (Wilks, 1967). It is true that in Ashanti there is no ruling estate with distinct traditions of origin and culture; in war the state always relied upon the divisional forces and the armies of tributaries. But successive monarchs founded additional regiments of musketeers (*asafo* or *fekuo*) in Kumasi; these companies offered an alternative force solely at the monarch's command, the membership of which mostly cross-cut the basic kin groups in the society since they were recruited by paternal filiation rather than by matrilineal descent.

In Gonja power was much more diffusely spread throughout what, on the Irish pattern, I have called an over-kingdom (Goody, 1967). While each division had its own gun-men, the rulers themselves were essentially horsemen, constituting a mass estate through which the major political offices of the realm circulated in a broadly distributive manner.

The same kind of difference, particularly the contrast between the modes of vertical and lateral succession to office, is brought out in Bradbury's illuminating comparison of primogeniture in Benin with the system of circulating succession that was practised in the Yoruba kingdom of Oyo. In Benin he sees 'a greater potential for monarchical autocracy than is found among the Yoruba. The rule of succession, the absence of large lineages with continuing rights in offices, and the open character of the palace association system, have given the Oba of Benin greater security of tenure and a greater freedom to manipulate political mechanisms than were available to his Yoruba counterparts. The absence of a powerful royal lineage giving backing to the Oba, on the other hand, might be thought to work in the opposite direction' (1964: 159). In contrast to autocratic Benin, Bradbury points to the wide-based lineages of the Yoruba, to the power of the Oyo Misi, who could unmake as well as make the king, to the close rather than the distant conflicts over succession. Among the Gonja, the electors have little of the importance of the Yoruba kingmakers nor is there any legitimate procedure for unmaking a king. But the general contrast with the vertical systems of succession still holds; in Gonja circulation is found in a mass dynasty which plays a major role in the nation's affairs.

The generality of this contrast should perhaps lead us to supplement Bradbury's hypothesis that the differences between the dynasties of Yoruba and Benin, which have a common origin, are in some degree 'explicable in terms of a process of adjustment to basic Edo social and cultural patterns' (p. 159). As with Lombard's examination of the Bariba and the Fon, the contrast is again one between savannah and forest states. These differences in ecological context had an important bearing on the situation; in the former the horse was the basis of military organization, in the latter the gun; with the horse one required a mass dynasty, with the gun one could dispense with all except a stem dynasty.

Many of the important differences between the forest and savannah states seem to relate to military organization. In Dahomey the monarch built up not only a central government but also a central military organization, of which the famous Amazons (conceived as 'wives' of the king) were but one example. Trade was under strictly centralized control.[8] In the savannah, commerce is largely independent of the government which simply provides the conditions under which it may flourish and takes a rake-off on the results. In the south the external trade was often nationalized, or at least heavily influenced by national policy.[9] The position of the royal dynasty was whittled away;

[8] Polanyi remarked upon this fact, which is discussed by Rosemary Arnold in Polanyi *et al.*, 1957. But the author does not differentiate between the exchange operations of forest and savannah states. Moreover the interesting discussion is given a peculiar twist by the insistence on the separation of trade and market, the latter being regarded as 'the transcending principle' (p. 186; see also Polanyi, 1966). But an economy which was so dependent upon overseas trade with advanced countries was perhaps less archaic than Polanyi suggests.

[9] In Benin, too, trade was highly centralized. Of a sculpture of a Bini trader with his staff of office, holding a manilla in his left hand, Dark writes 'Only certain people, such as this trader, held the king's licence to trade with Europeans. The Europeans, in turn, were allowed to trade only with the specially appointed Bini traders and were not allowed to trade in the native markets' (Forman and Dark 1960: 31).

its members filled no important posts; the armed forces were full of slaves.

What were the technological features that permitted the spectacular differences that Lombard, Bradbury, and others have noticed? Partly of course, the nature of the sea-borne trade with foreigners, which made for relatively easy state control, compared with the free-enterprise generated by the northern trade across the deserts and savannahs of the Western Sudan. But very important was one of the main objects of this trade, the gun itself. Unlike horses, guns are centralizable; they can be kept in magazines in the capital and issued when required for state purposes. The central government can maintain control of their distribution because of the relative complex process of manufacture. As we have noted, few, if any, guns were made in Africa, unlike Ceylon and Japan which quickly took over the manufacture of this Western invention.[10] The importation of these weapons from outside (and of gunpowder too) made central control even more feasible; this was also necessary in order to obtain the goods needed to buy guns, that is, the gold and slaves. In Ashanti the king had many royal rights over gold, while slaves were gotten either by tributary arrangements or by organized raiding parties, backed by the guns for which the slaves were exchanged.

It is unnecessary to labour the point that from the end of the seventeenth century the coastal states of West Africa depended heavily on the importation of European guns and gunpowder. Artistically, this emerges in the themes of Benin sculpture. The rise of Akwamu, Denkyira and Ashanti are directly connected with these events; so too is the position of Dahomey. At a yet earlier date the gun had made its appearance in the savannah country and Ibn Fartua reports the use of Turkish musketeers by Mai Idris Alooma of Bornu, who reigned at the end of the sixteenth century: the Turks dominated the North African scene after conquering Egypt in 1517. Along the coast the exception was Yoruba, where the gun was not important until the 1820s (Ajayi and Smith, 1964: 17).

[10] On the position in Ashanti, possibly the wealthiest and technologically most advanced state in West Africa, Kyerematen writes: 'Efforts were made to manufacture guns locally, but with little success, owing to the difficulty of obtaining iron ore; they resulted in the *humu* gun, made of brass, so called because of the barrel which was absurdly large' (1964: 39). Low-grade iron ore is widely available in Africa, but it was little exploited by the Ashanti themselves who relied on the importation of iron in exchange for kola.

This fact is important for my thesis since Oyo in the north was also a cavalry state. It also had a much 'wider' method of succession, a circulating system like a number of the savannah peoples. Why? Because while primogeniture is a mode of concentrating political power, circulation spreads it among the members of the ruling estate, who not only maintain their own position but also the military effectiveness of the state. When military effectiveness depends on the gun rather than the horse, the king can work with a much smaller dynasty depending upon slaves rather than equals. However, to do so is to sow the seeds of one's destruction. By placing one's trust in slaves, the way is opened for revolution as well as rebellion, a slave revolt of the kind that worried the Ashanti king when he talked with Dupuis,[11] that Clapperton reported from Ilorin,[12] and that usurped power in the Bambara state of Segou in the middle of the eighteenth century (1736–48; Monteil, 1924:102). This was like the situation of the Turkish monarchy under the Janissaries; the dynasty is so highly centralized that it lays itself open to a military take-over.

The military effects of the firearm require little emphasis, for they were world-wide. It is the theme of McNeill's study of Europe's steppe frontier: in the sixteenth and seventeenth centuries A.D., 'firearms, standing armies, and the supporting elements of modern civilised warfare reversed the age-old balance between steppe and south and drove the nomads into permanent retreat' (1964: 7). For weapons of this kind could be produced only in settled communities; in these the processes of production and the changes in military organization resulted in the greater centralization of government, 'the victory of monarchical bureaucratism' (p. 128). Similar repercussions were felt in West Africa, even with the few imported arms that were available. Ibn Fartua remarks of the first use of firearms by Mai Idris

[11] One Ashanti spoke to Hutchinson of the danger of slave revolts in Kumasi (Bowdich, 1819: 381). Bowdich, the leader of the mission, estimated the population of the town as over 30,000, which was supported by slaves in the bush. When Dupuis visited the country a few years later the King of Ashanti was quite explicit about the possibility that too many slaves might lead to a revolt (1824: 164).

[12] In his account of his second journey, Clapperton mentions a slave revolt in Yourriba and also the fear expressed by the rulers of Kano of attacks by slaves on their masters; there was said to be thirty slaves to every free man. 'It had been customary, in cases of this kind, to send the perpetrators of similar crimes to the sea-coast, to be sold to the slave-dealers' (1829: 171–2).

Alooma of Bornu, hitherto a cavalry state: 'Among the benefits which God (Most High) of His bounty and beneficence, generosity, and constancy conferred upon the Sultan was the acquisition of Turkish musketeers and numerous household slaves who became skilled in firing muskets. Hence the Sultan was able to kill the people of Amaska with muskets, and there was no need for other weapons, so that God gave him a great victory by reason of his superiority in arms' (1926: 11–12).

The sultan won 'by reason of his superiority of arms'. But these arms were in the hands of slaves and mercenaries. The ruling group stuck to their horses, a choice that would receive much support from the ideology of Islam. In this area horses were especially useful; cavalry could make a sudden descent upon the enemy and then retire to the safety of a stockade or walled town. '[The enemy] kept seizing the walled towns of the Yedi as fortresses and places of refuge and hiding, using them as bases treacherously to attack the Muslims by day and night, without ceasing or respite' (1926: 13). Mai Idris himself constructed such a base. 'He therefore built the big town near Damasak. . . . He made four gates in the town and placed a keeper in charge of each gate and quartered there a detachment of his army. He ordered all his chiefs who were powerful and possessed of a defence force, to build houses, and leave part of their equipment there as for instance, the horses, and quilted-armour for them and coats of mail' (1926: 15). Cavalry warfare and raiding still represented the basic military activity of the savannah states; the guns attained much less continuing importance. As in the case of Samory, their use could be very effective, but distance from the coast of West and North Africa meant that problems of supply were always critical and much greater than that of a largely self-maintaining cavalry, though even horses too had often to be imported from outside.

Two broad types of polity have been distinguished in pre-colonial Africa, the acephalous and the centralized; these the Ghanaian G. E. Ferguson (who was much concerned with the pragmatic aspects of the distinction for colonial policy) called 'barbarous tribes' and 'organized governments'. Each type has many variants; indeed for some purposes they can be considered the poles of a continuum. But looked at in terms of the control of force (and many definitions of politics and law turn on just this point), they are distinguished by the military technology available to them. The distinction is not hard and fast, states can arise on the basis of the technology available to acephalous peoples. But by and large there is a considerable measure of correspondence.

Acephalous societies have a mainly bow and arrow culture; states are largely based on firearms or the horse. Though some firearms were found in the savannah zone, it was the coastal states that depended upon the gun; the grassland rulers are the horsemen.

Indeed the states of Ashanti and Dahomey refused to allow guns and powder to pass through their territories to the inland kingdoms; for it was precisely their control of these weapons that enabled them to dominate the interior and to extract from its peoples the slaves they needed to purchase more guns and to maintain their standards of living.

The dichotomy between the gun states of the forest (where the horse could not operate, partly for trees, partly for tsetse) and the horse states of the savannahs corresponded to significant differences in the social systems.

In the grasslands, the ruling estates tended to be mass dynasties, within segments of which high office often circulated. In the forest, office tended to be more autocratic and to be retained within a narrower dynasty. In the former, power was more widely diffused and the rulers relied on the dynasty as fighting cavalry; in the latter, the important fighting arm was the regiments of gun-men stationed around the capital and often recruited from prisoners and slaves.

Other differences were connected in a less direct way. Since guns and gunpowder came from the sea-borne trade, it was easier for the state to control the local dealers in arms. Not only were the kingdoms more centralized as a result but a large sector of the external trade was state operated. In contrast, the markets and entrepôts of the interior were dominated by the free enterprise activities of Muslim merchants. Indeed the distinction between the Islamic traders of the interior and the Christian traders of the coast was another important aspect of the contrast and one which had significant socio-cultural implications. This situation was part of the world-wide confrontation brought about by the development of firearms. Within both Western and Muslim states the internal changes brought about by this invention were far-reaching. But as Ayalon pointed out, they had a profound influence on the course of the struggle between these two worlds. 'By and large. . . the introduction of these new weapons spelt disaster to Islam' (1956: ix); the outstanding exception was the success with which Ottoman Turks used firearms against both their Christian and Muslim neighbours.

For social organization, the most significant aspect of distinction between acephalous societies, forest kingdoms, and savannah states lay in their military technologies rather than their productive techniques. In the broader sense the control of force had its more productive side, at least for the ruling estate; and the availability of military goods (the gun and the horse) depended upon a system of exchange that extended as far as the Baltic for guns and the Barbary Coast for horses. But in terms of social organization, one major factor in the distinction between acephalous societies, forest kingdoms, and savannah states lay in differences in the means of destruction and in their ownership. It is because the crucial control of force was the first element of the systems to disappear under the colonial regime that this factor has so often been underplayed in retrospective analyses and post-colonial reconstructions.

## CHAPTER 4

# POLITY AND RITUAL: THE OPPOSITION OF HORSE AND EARTH

THE differences in West Africa between the military technologies of European nations, forest kingdoms, savannah states and the acephalous peoples were of major importance in determining the power relationships between them. For one is dealing not with so many isolated tribes or nations, but with a set of interacting relationships. For example, the Ashanti desire to acquire slaves for the European trade led to pressure upon the tributary states of the north, especially Gonja, to produce the human booty. This in turn led to cavalry raids on the horseless, acephalous peoples such as the ' Grunshi ', the Konkomba, the LoDagaa and the Tallensi. Because such acephalous peoples were regarded as pools of manpower and could do little to resist the incursions of their centralized neighbours, they tended to occupy land which was difficult of access, especially to horses. It is not accidental that many of the acephalous peoples are found either in hill areas, like the Dogon of Mali, the Tallensi of Ghana, the inhabitants of the Togo hills or the Bauchi plateau, or are straggled across a major river, like the LoDagaa and the Konkomba. A neglect of these interactions between state systems and acephalous peoples has impoverished our analysis of the total social situation, much as if one were to analyse the Zulu without the Whites.[1] A complex network of the kind revealed by travellers' accounts (e.g. Binger, 1892) and in the writings of the indigenes (Braimah, 1967; Krause, 1928) was less visible to those whose information was gathered under the umbrella of colonial or post-colonial rule. But it must clearly have had a major place in the ' world-view ' of these peoples; one cannot lightly leave aside devastation of the kind that marked Babatu's establishment of a booty enclave in the centre of Grunshiland, nor the descent of Samory's forces upon the LoDagaa or western Gonja. Myths,

[1] On this point, see Gluckman, 1940.

cosmologies, systems of thought were inevitably influenced by such radical and recurring events. One example of this is to be found in the ritual and symbolism that has accreted around the horse and the earth in the Western Sudan. In order to point to the relationship between ritual and war, I turn to evidence that comes from one of the earliest confrontations between European and African in northern Ghana, at a time when relative force was a critical determinant of social relations.

In the last decade of the nineteenth century armed bands of European invaders marched through much of the savannah country north of Togo and Dahomey. They were able to roam more or less at will because, while they found themselves in possession of automatic rifles, the local inhabitants had at best muzzle-loaders (or 'Dane guns' as they were known locally) and at worst spears, swords, and bows: and even the muzzle-loaders and the best powder had to be obtained eventually from enemy sources.

These small bands of marauders, accompanied by slaves and servants, and firmly clutching their automatic weapons, represented the major powers of Western Europe. The hinterland of the colony of the Gold Coast, with which I am mainly concerned, was mainly the preserve of the British, though quite how far back their notional territory extended remained vague. The French were rapidly approaching this area from the north-west, and the Germans were cutting in from their base in Togo, anxious to make up for their late start in the scramble for Africa. Just because they had made so late a start, they were prone to see themselves at a disadvantage against their imperialist rivals, England and France, who had already acquired their additional *lebensraum*. Hence they tended to be more militant in their methods. Their carriers as well as their soldiers were armed with guns; this, and the competitive pressure under which they worked, meant that though they seem to have been better educated, more learned, than their French, and certainly their British, counterparts, they were also more prone to incidents of a violent nature.

The leader of one such band was Hauptmann Kling, who, early in 1893, visited the Gonja town of Salaga, the important market in the kola trade between Ashanti and northern Nigeria.[2] From Salaga he passed down to the town of Kintampo. For, following the British entry into Kumasi in 1874, eastern Gonja had freed itself from Ashanti dominion and massacred the traders who were the local representatives of that powerful empire. As a result, the Ashanti had switched their kola trade to the town of Kintampo in northern Nkoranza. However, at the time of Kling's visit to Kintampo, Nkoranza too was involved in a revolt against Ashanti and the road was once again blocked for the kola trade.[3]

On his return from Kintampo, Kling passed through central Gonja, visiting the ancient trading town of Buipe (or Gbiipe) which had a largely Muslim population. He then approached the town of Bute, which was inhabited by members of the commoner estate alone; the chiefs and Muslims lived in other villages in the vicinity. This town, with its three-storeyed tower-houses and its many cisterns, had played an important part in resistance to the Ashanti. Shortly afterwards Bute was a focus of similar resistance to the forces of the Mandingo warrior, Samory Turay, then trying to make contact with the Germans at Krachi in order to sell slaves and obtain arms.[4] The cosmopolitan trading town of Buipe had no such defences at that time and its population fled before the Mandingo advance. When the British reached there in 1897, it was still completely deserted.

Kling was dissuaded from approaching Bute; no one would even show him the way 'because the chief doesn't allow any white or Ashanti man in his village'. As he came near, his party were threatened by men armed with Dane guns; it remained under their surveillance until it had passed by on the track to Salaga.

The refusal to allow Kling to enter Bute was based partly upon religious sanctions, for this village was the home of an important shrine brought up from the Longero region in northern Nkoranza at the time of the Akan expansion in the early eighteenth

[2] See Goody and Mustapha, 1967; Braimah and Goody, 1967.

[3] Kling notes that the Ashanti had officially recalled their traders from Kintampo, but many of these had refused to leave.

[4] Goody, 1965.

century.[5] Like many other shrines in the area, one of its main functions was to protect its clients against misfortunes, in this case, armed attack. But as well as this ritual sanction, the town was at once a sanctuary and a defensive stronghold, because of its storeyed buildings constructed close together on artificial mounds.

Before he reached the White Volta, which lies some miles east of Bute, Kling came to the village of Yaurupe, where one path led to Yendi, capital of the Dagomba kingdom, and another to Salaga. Here again he ran into local taboos and his chronicler writes 'Some closed grass huts are consecrated to a fetish, which can't tolerate horses, so the traveller had to leave his horse somewhere far away'. He crossed the White Volta, was headed south by some armed men, probably belonging to forces of the Gonja chief of Kusawgu, and passed through Jukuku to another small village, where horses were again taboo. Once more Kling's mount had to be led around the village in order to prevent it from being killed by the local shrine.

These instances from central Gonja are by no means the only examples we know of ritual prohibitions against horses. In 1951 the District Commissioner at Bole in western Gonja, Tomlinson, records that when he rode to the 'fetish' village of Senyon he was asked to leave his horse outside 'because there is a tree which no horse may see and survive' (1954: 9). Some fifty years earlier, the first British forces occupied this area of Gonja. On 7 February 1900, a group of African soldiers, under the command of a European officer, Lieutenant Roy, arrived from Bole and asked the chief of Senyon to buy food for them

[5] Now inhabited by Grusi-speaking Dega (or Mo) who were tributaries to the Ashantihene through Nkoranza, Longero was formerly under Gonja sovereignty (see the *Gonja Chronicle*, Goody, 1954: 38). It was attacked by the Akan in the early eighteenth century and the Gonja withdrew to the northward bend which the Black Volta makes above the 'Desert of Ghofan' (Bowdich, 1819; Cofie, 1963). The main towns of the Gonja kingdom later withdrew yet further across the river. The custodian of the important shrine of Elampo then went to Benyalipe, and an altar was also set up at Bute. The Ashanti from Nkoranza (the northernmost division of their realm) established themselves at Kintampo, with a permanent outpost at Dawadawa, along the road to central Gonja. It was from here that, at the beginning of the nineteenth century, the Ashanti fought a series of battles against the Gonja who were in alliance with the Abron (Jaman) kingdom of Bonduku; these battles are reported in various European sources (see Goody, 1965: 29 ff.), and are connected with the attacks against Bute remembered in Nkoranza tradition.

and also to pay certain arrears of tax. The chief was unable to find the money, and the food he produced was deemed insufficient; as a result the senior divisional chief in Gonja, a descendant of the eldest son of the founding ancestor of the state, Ndewura Jakpa, was made to spend the night in the soldiers' camp. Next day the chief was asked to provide two carriers. When none arrived, he was told to pick up one of the loads himself. Lieutenant Roy then sent off his party of carriers, accompanied by three soldiers, two of them mounted. In order to get on the right track they had to pass through the middle of this tight, flat-roofed town. As they were going through, a drum sounded and all the able-bodied men appeared on the roof-tops, levelling a mixed bag of weapons at the soldiers in the alleys below. When Roy arrived on the scene, he found the soldiers defending themselves against the inhabitants who were trying to pull them from their horses. As a result of the mêlée the chief escaped and sent a message ' saying that he was tired of " the white man ", that he could not pay any money, and that if his soldiers came to or through his town he would kill them, that Samory could not touch him, that therefore white men could do nothing '.

One-sided as it is, the official account has much of interest; for the attack was aimed not so much against the soldiers or their officer, despite the ill-treatment they had meted out to a senior Gonja chief, but rather against the horsemen who had dared to ride through a town that contained the most important commoner shrine in Gonja, Senyon Kupo, and had very close associations with the earth in its mystical aspect. Like other such shrines, Senyon is not in the custody of the ruling Gonja, descendants of invading cavalry (though these live in the same town), but rather of a shrine priest (*kagbirwura*) who claims an autochthonous status, at least relative to the Gonja. So great was its reputation that even the forces of Samory Turay, who made an important base at Bole some dozen miles away, never descended upon the town. They spared these ' pagans ' even though their Muslim co-religionists of Bole they had used as carriers, while the Jegbagte believers of nearby Dokrupe, the Muslims of the Senyon division, they drove south of the Black Volta. The European invaders were clearly placed in the same category as Samory's men. These latest newcomers were, however, less concerned with the supernatural powers, less knowledgeable too of local beliefs. For this ignorance the Europeans paid a price.

As a result of this local setback, the acting commandant of the Northern Territories' Force, Major Morris, intended to visit Senyon himself but was called on other duties. Seventeen years later, when the colonial powers were engaged in an internecine conflict among themselves, a conflict which put increasing pressure upon their subject peoples, Eastern Gonja rose against the white man, like many other parts of West Africa where 1917 was the year of revolt.[6] In Gonja the centre of revolt was the same town of Senyon, which provided the rallying point for local resistance. The result was never in doubt; apart from burning the Senyon rest-house, little violence took place. At the approach of the British garrison under Colonel Moutray-Read, resistance disappeared; the Gonja soon realized they were no match for the colonial machine-guns, of whose power they had constantly been made aware by 'peaceful' demonstrations.[7] But still today, horses are taboo at Senyon; so too are any close relationships with Europeans; on a recent visit, when I proffered the priest my right hand, his assistant (*kupo*) informed me that the shaking of hands (a Muslim as well as a Christian gesture) was forbidden by the shrine.[8]

Another and rather similar situation existed further eastwards at the town of Krachi, abode (until the recent construction of the Volta dam) of the god Odente. This shrine was found near the junction of the river Dakar and the Volta, at a point just below the rapids that interrupt all river transport to the north. Here

[6] The Bobo risings started in the previous year (Capron, 1965: 130); other risings occurred among the Tuareg in 1917 (Miner, 1953: 13), the Mossi and in northern Togo. This was also the beginning of the mass migration of Lobi (LoBirifor) over the western boundaries of northern Ghana, and of less extensive migrations into the 'Fra-fra' region north of Bolgatanga. There were riots in Fra-fra country too during the same period, where the idea of the departure of the white man, involved as he was in intra-racial war, was encouraged by the withdrawal of the garrison from Zuarungu.

[7] Kling records how he destroyed a wall with gunfire in order to demonstrate the power of his weapon. Such demonstrations were part of the stock-in-trade of colonial rule, as of other conquest states. 'Showing the flag' usually meant a public display of the means of destruction as a deterrent to local action.

[8] In medieval Europe, too, shaking hands had a more than casual significance; the hand-grasp was a means of sealing a contractual relationship (Pollock and Maitland, *The History of English Law*, 1952, ii, 188).

it was that Ada traders from the mouth of the Volta, taking salt and manufactured goods to Salaga, had to off-load their canoes and proceed by land through Bajemeso, or else porter their boats round the rapids and continue up to Yeji. This strategic town of Krachi, that dominated the trade route down the Volta valley, was conquered first by the Nanumba, probably by the Gonja, then by Ashanti from Juaben, and finally by the British and Germans at a time when the decline of Salaga had led to the great expansion of Kete, the twin Muslim town to Krachi and the eastern counterpart to Kintampo in the west.

Like Senyon Kupo and Burukũ in nearby Shiare, the Odente shrine at Krachi was widely patronized, being regularly consulted by the Chief of Ashanti, especially when the Europeans threatened to invade his country. Once again this shrine was vested in the commoners, the people of the land (as opposed to the local invaders like the Nanumba, Gonja or Ashanti), and it was also anti-horse. One of the first German administrators in the area wrote: 'Besides the goats the Odente hates . . . donkeys and horses. For this reason they made a special caravan track for these animals outside the town because they did not want to forgo the profitable ferry business' (Klose, 1899: 341).[9]

There are many other instances of shrines under the custodianship of autochthones having taboos against horses. One striking instance is the famous hill shrine of the Tallensi in the Tong Hills. This widely known shrine was in the charge of the indigenous Talis whose position is often opposed to the immigrant Namoos, an offshoot of the ruling estate of Mamprusi.

The association between shrines, autochthones and anti-horse taboos is a feature of what Fortes called the major cleavage among the Tallensi, between Namoos and Talis. The opposition between the two, which has been formalized over time, emerges at important funerals when the autochthones chant against the immigrants: 'we might rise one day at break of dawn to sally forth against certain people' (Fortes, 1945: 27). The Talis are the 'indigenous' agriculturists, linked with the earth and its shrines: the Namoos (in tradition at least) are off-shoots of the horsed invaders who established the Mossi-Dagomba states.

[9] According to Klose, Burukũ was consulted by the kings of Ashanti, Dagomba, Salaga and Tshautsho: 'they all come before an impending war or feud, to have its results prophesied' (1899: 439).

It was the horses and guns of the immigrant leader, Mosuor, that inspired fear in the hearts of the indigenous Tali Earth priests and led them to accept these strangers in their midst (1945: 22). The opposition between the two groups takes many visible forms, the red fez and cloth tunic are the robes of chieftainship, whereas the Earth priests (*tendaanas*) must wear skins, cloth being forbidden to them since it is widely associated with status. But the relationship emerges at its clearest in the fact that ' chiefs and their clansmen may ride horses, *tendaanas* not '.

In the Tallensi case, the Mamprusi immigrants have in the course of time lost any dominant position they may originally have had and reached a kind of parallel status with the indigenes. Elsewhere, within the effective boundaries or spheres of influence of the states, shrines (especially those of the earth) are usually found in the hands of the autochthones while the invading cavalry groups possess the means of their subjugation. The more pragmatic orientation of the conquerors made them interested in terrestrial rather than supernatural power. Many of their connexions were with Islam (Rattray, 1932: xii); the major state ceremonies are Muslim festivals. But both by tradition (for indigenous religions are singularly eclectic) and by policy (for chiefs had to rule pagans as well as Muslims), the religious practices of the ruling estate were certainly not confined to one particular mode of supernatural communication. They were usually prepared to recognize the superior wisdom of the autochthones about the things of the Earth, and of Muslims about the things of God.

The situation in northern Ghana is duplicated in other parts of the Western Sudan. Indeed when Sir James Frazer was trying to illuminate the relationships of conquered and conqueror that centred upon the Earth, it was to the Mossi, the northernmost of the Mossi-Dagomba states, that he turned. Frazer was looking for parallels to a biblical incident upon which Robertson Smith had already commented when dealing with the importance of local shrines in the Middle East, the *Baal* of the early Semites. ' The Aramaeans and Babylonians whom the king of Assyria planted in northern Israel brought their own gods with them, but when they were attacked by lions they felt that they must call in the aid of " the god of the land ", who, we must infer, had in his own region power over beasts as well as men ' (Smith,

1889: 92). Thus the autochthones, weaker in military might, were stronger in supernatural power, which served as some counterweight to their political inferiority. It is in this context that, in a chapter in *Folklore in the Old Testament* entitled ' Jehovah and the Lions ', Frazer calls attention to the very similar situation that Louis Tauxier describes among the Mossi: ' The existence of these Chiefs of the Earth among the Mossi is explained very probably by the superposition of the conquering on the conquered race. When the Mossi invaded and conquered the country, in proportion as they spread their dominion they put men of their own race at the head of all the villages and cantons to ensure the submission of the vanquished population. But they never thought—and this is a notion to be found in the whole of West Africa—that they were qualified to offer sacrifices to the Earth-god of the place and the local divinities. It was only the vanquished, the ancient owners of the soil, with which they continued in good relations, who were qualified for that. Hence the old political head of the aborigines was bound to become naturally a religious chief under the rule of the Mossi. Thus we have seen that the king (*Moro-Naba*) never himself offers the sacrifices to Earth at Wagadugu, nor does he allow such sacrifices to be offered by his minister of religion, the Gandé-Naba. He lays the duty on the king of Wagadugu (*Wagadugu-Naba*), the grandson of the aborigines, who as such is viewed favourably by the local divinities. Similarly, when he sacrifices to the little rising-grounds in the neighbourhood of Wagadugu, he commits the charge of the offerings and sacrifices to the local chief. But what the king (*Moro-Naba*) actually does now at Wagadugu, the Mossi kings (*naba*) doubtless did formerly, more or less everywhere after the conquest, as soon as the submission of the aborigines was assured. Hence the institution of the Chiefs of the Earth (*Tensoba*) ' (Frazer, 1919: iii, 86–87).

The Senyon shrine also had a close connexion with the earth. Like the Tongo of the Tallensi, the Tigari of the Nome (Ypala, Gonja), the Kungpenbie shrine of the LoWiili of Birifu, Senyon Kupo had an extensive influence in southern Ghana in the 1930s, and possibly before. M. J. Field came across the shrine in Akim-Kotoku and subsequently went to visit the priest at Senyon. Her report of his speech emphasizes the close connexion with the Earth. ' They said, " *Kupo* has no house except the Earth. He lives inside the earth and helps the earth to grow crops. No one has ever seen him, but he goes everywhere. He is a god of the soil and we give him of our crops for his food. If you belong to him you must not lie or steal or he will kill you" ' (1948: 181). It is this same shrine, here shown as closely identified with the Earth, which as we saw before is so strongly opposed to the horse.

The widespread linkage of autochthones with taboos against the horse is not accidental. Throughout the whole of the savannah regions of the Western Sudan, hoe agriculturists, armed with the bow and arrow stood opposed to ruling estates, whose dominance was based primarily on cavalry; some simple guns were owned by members of both groups, but owing to difficulties in the supply of guns and powder discussed in the previous chapter, only rarely were they the dominant weapon in warfare. The commoners of the savannahs, whether they lived in stateless societies or under the umbrella of a state system, were warriors with the bow and arrow, the weapons that they used both for hunting and for war. But the horses were owned principally by members of the chiefly estate, who had established their original domination by their use and who maintained their position, internally and externally, by raiding for slaves and by taxing the traders.

In Africa south of the Sahara the horse was never used for agricultural work, since there was no plough to pull, no wheeled vehicles to draw. It was the humbler donkey that served as pack-animal, the 'noble' horse being used for transporting high-status humans. Though used to carry merchants engaged in long-distance trade, the return on horse-ownership came mainly from its role in domination. Unlike the bow and arrow, which was employed as a means of production as well as of destruction, the horse could be used only for war, for travelling, and for the prestige that it carried.

As in the case of the commoners, the association with the horse took on ritual significance. The annual Damba (Ar. *mūlūd*) ceremony, carried out at the divisional capitals in Gonja, is for the Muslims a celebration of the prophet's birth and for the rest of the country an act of obeisance to the chief. At this time members of the ruling state turn out by night to perform a dance which explicitly imitates the movement of horsemen. At this time they are both horse and rider. The image is yet more explicit in certain speech forms used to refer only to the paramount chief of Gonja. At times he is spoken of as *bange* (horse) in such phrases as '*bang dɛsɛ*?' (Is the king well?). Once again horse and horsemen are identified.

In Eastern Gonja every horse belonging to a chief had assigned to it one of the chief's wives and a grass-boy. The wife was responsible for bringing water to the horses which, like fetching grass, often entailed a long walk. Three times a day she offered the horse water that had been mixed with flour. As she did so, she dropped to her knees and agitated the liquid with her right hand so that it was well mixed (E. N. Goody, field-notes, 1964).

The same identification of horse and rider occurs throughout the savannah kingdoms, many of whom claim a common origin. Thus the Mossi-Dagomba states all trace their origin to one founding ancestor and their common legends link them both with the Hausa (and through them with Bornu and the Berbers) and with the Mande group of states. The founding ancestor of the Mossi kingdom of Wagadugu was again known as 'the red stallion' and the same surname of Wedroago is still common today among Mossi of the ruling estate.

The identification is reinforced by contact with Islam, with which most of these savannah states of West Africa were associated in some way or another. The cultural tradition of Islam emphasizes the life and values of horse and camel nomads. While the centre of Islam was a market place, it was the life of the tented Bedouin that formed the back-cloth of its literary activity. For example, the classical *qaṣīda* or Ode '. . . must begin with the mention of women and the constantly shifted habitations of the wandering tribesmen seeking pasture throughout the Winter and Spring. . . . From this theme, he turns to the main object of the poem, either abruptly or by interposing the description of his horse and camel, by means of which he escapes from the burden of memory when it grows too hard for him . . .'[10]

This tradition accurately reflects the course of Arab history, since it was partly because of the superior mobility and military

[10] Charles James Lyall, *Ancient Arabian Poetry*, London, 1930, p. xix f., quoted Hiskett, 1963: 9. The continued concern with the nomadic life is very strong in Ibn Khaldun's sociological analysis, *The Prolegomena* to his *Universal History* (Issawi, 1950).

advantages of their camels and horses that in the seventh century A.D. the Muslims were so quickly able to establish their rule throughout North Africa. And it was often through this control of cavalry that they were able to maintain and extend their dominion yet further. We must therefore see the success of the Arabs and Islam in a wider context, for they were but one in a line of conquerors whose superiority was based upon the use of the domesticated animals, especially the bovines, horses, and camels, which penetrated the Sahara in just this order (Lhote, 1960: 191). As far as Africa is concerned, none of these animals was domesticated there. Indeed the wild horse was confined to the northern steppes of Europe; in Africa the ass was found along the Mediterranean shore and the zebra in the eastern savannahs, but only the first was ever brought under direct human control. Although Africans south of the Sahara domesticated their own wild plants, they domesticated none of their wild animals.

The horse itself was originally domesticated in central Asia and became the main instrument by which Indo-European-speaking peoples such as the Hittites extended their dominion into the Near East. About 1500 B.C. the Hyksos invaded Egypt and it was from them that Egyptians took over the chariot. In the course of the first millennium B.C. the use of these horse-drawn vehicles spread into the Eastern Sudan and across the Western Sahara. It seems likely that here as elsewhere the owners of these animals were able to establish their political supremacy by virtue of their weaponry and of the mobility that the horse gave them.

Be this as it may, the first states that we hear about in the savannah zones of Africa are linked with horse-owning dynasties. Even the small amount of information available about the earliest known kingdom of the Western Sudan brings out the importance of the horse. Describing the court of the King of Ghana, Al Bakri (A.D. 1028–94) wrote: 'The court of appeal is held in a domed pavilion around which stand ten horses with gold embroidered trappings.' Even at this time the horse was clearly the noble animal, associated with the ruling estate, while the transport of goods was left to the donkey, owned by the merchants. 'From every donkey loaded with salt that enters the country, the king takes a duty of one golden dinar, and two dinars from every one that leaves.' Part of the reason for this difference was undoubtedly the role played by the horse in military action, and it is significant that Al Bakri's final mention of the horse is made in the context of army organization: 'When the king of Ghana calls up his army, he can put 200,000 men into the field, more than 40,000 of whom are bowmen. The horses in Ghana are very small.' The implied contrast in size was with the larger Arab horses; these were animals of the smaller breeds found widely distributed in the savannah country.[11] Later accounts of this kingdom also stress the role of the horse. In the famous geography of Idrisi (A.D. 1099–1154) a reference is made to the great nugget of gold to which the king tethers his horse. Later we are told how the monarch rides round the town twice a day, at the head of his principal captains, to investigate and remedy the injustices or misfortunes suffered by his subjects. And finally, in the third reference to these animals in a fairly short passage, the king is said to ride 'only on horses'.

Thus like their neighbours to the north, the Garamantes of the Fezzan mentioned by Herodotus, their descendants the veiled Berber-speaking Tuareg, and the later Arab tribes that penetrated the Sahara, the ruling dynasty of Ghana were closely associated with 'military' livestock. Indeed there is even a suggestion that a modified form of the 'white-black' stratification existed in ancient Ghana, as among present-day Tuareg, consisting of 'white' horsed rulers and 'black' horseless cultivators, though the first reference to the 'white' northern kings of Ghana does not occur until long after the kingdom had disappeared.[12] When it did collapse, it fell to the Berber of the Almoravid sect from Senegal, whose cavalry long dominated the trans-Saharan routes.

Subsequent kingdoms in the Western Sudan all had a similar military base, Melle, Bornu, Songhai, Hausa, Bariba, Gurma, Gonja, and the various Mossi-Dagomba kingdoms of the Voltaic region. Indeed it has been said of this area that 'states arose on the backs of horses'.[13] The same source suggests that the relative antiquity of the horse in the Western Sudan can be gauged by the 14 root forms of the word in the Gur languages alone,

[11] Quotations from Al Bakri, *Description of Northern Africa*, translated in Fage, 1957, esp. pp. 81–2.

[12] In Ta'rīkh al-fattāsh, by Mahmūd Kāti (d. 1593).

[13] Köhler, 1953–4.

though this fact may derive partly from the variety of types of horse present in the region and partly from the attention devoted to them.

From one standpoint, then, the Islamic movements of the nineteenth century, whether their accession to power was permanent (as with the Fulani in Hausaland) or temporary (as with Babatu and Samory), were in a real sense a continuation of an earlier and long-established tradition of raiding, conquest and state-formation that characterized these savannah regions. The question of the degree of Islamic influence in the states must be seen within this context. I do not wish to suggest that the ideological factors are insignificant in warfare. Any Islamic state, like its Christian counterpart, associated the use of violence with religious belief; it is God who legitimizes the breach of the prohibition on killing one's fellow men, for it is he who has sanctified that taboo. Moreover, these literate religions (and the exclusive, world religions are all of this kind) even provide for a justifiable opening for warfare against one's co-religionists, for it is always possible to show that any opposition group is failing to conduct its affairs according to the Book. In the savannahs of the Western Sudan, conquest was often associated with Islamic reform. A ruling house, seen to be attached to the things of this world is open to a puritanical movement on its flank, as was the case with the Fulani *jiḥad* of 1806.[14] The chronicler of the famous Mai Idris of Bornu, blessed with the advent of Turkish musketeers, referred to his Islamic opponents as pagans and

[14] There was a doctrinal difference on this point between ʻAbdullāh, brother of Shehu ʻUthmān (dan Fodio) and Sultan Bello, son of the Shehu. The estrangement between them was basically due to the succession; Bello was created Sarkin Musulmi at the death of his father in 1817, although ʻAbdullāh was the Shehu's *wazīr* at the time and hence ʻ the senior of the Fulani lieutenants ʼ (Hiskett, 1963: 16). When ʻAbdullāh came to attend his brother's funeral, he found the gates of Sokoto shut against him; for, it was later explained to him, ʻ had he been present they could not have appointed anyone other than him as Sarkin Musulmi and this would have meant that his descendants and not those of the Shehu would have succeeded. This certainly would have given rise to strife within the Muslim community ʼ (p. 20). But there was also a disagreement on the attitude to be adopted towards rebels such as ʻAbd al-Salām, a Hausa mallam who had helped the Fulani conquerors and then refused to recognize allegiance to his overlord. Following al-Maghīlī, the Shehu and his son took the position that those who assist infidels are themselves infidels, but ʻAbdullāh held that a Muslim did not become an infidel unless he acted in furtherance of an infidel religion (p. 18).

attacked them not only for the booty but for their religious failings, for example, because chiefs judged, not the '*ulema*, the men learned in Islam.[15]

Looked at in a wider perspective, it was military superiority based upon the horse that enabled those conquests to take place. But to understand the situation we need to consider some of the wider problems of a cavalry-based state in the Western Sudan. In each of the states, provision for a supply of horses had to be made either by breeding or by purchase. Good horses were always objects of the import trade, because of the high value placed upon animals of Arabian origin that came across the Sahara. But even within the Western Sudan, the breeding areas were limited in number because of the distribution of tsetse, so that many states had to import horses from outside.

In northern Ghana, the Gambarga scarp, which lies in the state of Mamprusi, is the only area known to me where breeding took place in recent times, presumably because it was relatively free of tsetse. It is perhaps significant that this is the area from which the ' Mossi ' kingdoms (Wa, Dagomba, Nanumba, Gurma, the Mossi states) claim to have dispersed and the place where scraps of mail have been discovered in an archaeological context (Carter, 1964). In other parts of northern Ghana, all horses had to be imported, certainly in recent times. In Gonja I could find no record of any mare that had successfully foaled. All had to be purchased from traders or received as gifts. The acquisition of horses must have been on a large scale since their life-span was short. In 1957 the chief of Kusawgu told me that he had possessed 20 horses during his reign of some 20 years; if true, this total represents a very considerable investment in these prestigious and expensive animals, all brought down by horse-copers from the north.

An interesting comment upon the manner in which part at least of this trade was conducted is provided by a nineteenth century Hausa manuscript collected by the German traveller, Gottlob Adolf Krause (1928). The account centres around the activities of Babatu who was a native of Songhai (a ' Zaberima '), an area known for the breeding of horses, being relatively free

[15] For example, the people of Kano are treated as pagans during these wars (1571–83), yet they had long since received Islam (Ibn Fartua, 1926: 6, 30). The people of Bornu are identified with ' the Muslims '.

of tsetse fly. Babatu led a group of adventurers in the Grunshi area, who plundered the country lying between the various Mossi-Dagomba states,[16] lived off the land and established a political organization that might in time have developed into a state (1928: 30. 5, 49–56). But it is clear that this leader first came to the Dagomba area as a horse-coper. The Dagomba chiefs promised to pay for the animals, so Babatu and his comrades decided to wait for their money. In the meantime they collected part of their debt by acting first as collectors of tribute, then as mercenaries and finally as freebooters. Throughout these various transformations, their position as horse-traders and horse-riders was the critical factor.

The difficulties of breeding horses, and indeed of keeping them alive, meant that arrangements had to be made for a continuous input of animals into many of the states. Goods or services had to be produced to pay for these imports. These goods were often slaves, the product of cavalry warfare, and were usually obtained by raiding the acephalous peoples, who had neither horses nor horsemen with which to counter-attack. The horses were then used to capture more slaves to pay for more horses; this human booty was the men of the Earth.

In conclusion, in the Western Sudan, differentiation in the means of agricultural production was not of major significance. Much more important was the ownership of the means of destruction, since on this depended political overlordship and the production of booty. And in the savannah regions such supremacy turned upon the horse. The commoners, then, opposed the horse because it was the instrument of their oppression. The rulers identified themselves with it because it was the instrument of their domination.

[16] Binger gives an account of a Mossi slave-raiding expedition that he encountered travelling through the region in 1888. It was led by Boukary, the future ruler of the Mossi kingdom of Wagadugu, who was heavily involved in raiding the acephalous peoples in the border areas of the state (1892: i, 470).

## CHAPTER 5

# CONCLUSIONS

IN a brief compass and in a highly generalized way I have tried to indicate some of the economic and political implications of the basic technological differences between Africa and Eurasia, as well as within Africa itself. The analysis of a political system needs to be related to its economic possibilities and these in turn are linked to the technology. On the productive level the technological gap was between a shifting agriculture based on the hoe, with iron a scarce and expensive commodity, and an advanced agriculture based upon the plough. One of the implications of this difference between Africa and Eurasia (reinforced by the relatively poor soil and the sparseness of population) lay in land tenure, that is, in the ownership of the means of production. These differences had important consequences on the political and economic level, so that many terms originating in the analysis of pre-industrial Europe, such as 'feudal', are inappropriate for the examination of the traditional state in Africa.

A recognition of these facts is not only important for historians and for a whole range of cross-cultural studies. They also bear directly upon current problems. The economic rights of the African rulers over land meant less than those of their European counterparts. When their military power was destroyed by the European conquests of the latter part of the nineteenth century, they had little to fall back upon except ritual status, ethnic loyalty and collaboration with the new dispensation. Except in a few situations, like that of Buganda[1] where the colonial rulers introduced a system of landlordism, chiefs did not directly control the usufructuary rights in land. Indeed in most regions, land had little scarcity value and was in any case *extra commercium*. The extension of cash crops and market production, which the penetration of European commercial and political activity brought with it, combined with the introduction of technical advances, increased the value and scarcity of land. A successful Ashanti cocoa farmer could bring a much larger acreage under use than a

[1] On the history of land tenure in Buganda, see H. West, *Land Proprietary Structure in Buganda* (African Studies Series, Cambridge, 1971).

hoe farmer, and he required yet further land as a reserve. Many of the difficulties centring upon chiefship in Ashanti had to do with the differences in the interpretation of the 'political' rights in land held by the chief and the exclusive farming rights that entrepreneurs wished to acquire for growing cocoa. As Busia points out when disputing Rattray's application of the concepts of English feudalism to Ashanti land tenure, the chief was the custodian of the land and his rights over any parcel of it coexisted with a cluster of other rights held by the lineages residing there. Each lineage had the right to use a particular portion and none could be sold without the approval of its members, living and dead. The chief himself was rich not in land, but in goods and services, which were not all for his own benefit. 'When a chief has plenty of breast milk', runs the Ashanti proverb, 'it is the people who drink it.' But since the advent of colonialism meant that the services ceased to come in while the goods continued to go out, the chief turned to private accumulation through trade or farming, leading to the necessity of drawing a more radical distinction between private and state property; formerly the chief's property had been state property (1951: 204). Under changing conditions of this kind, the position of the chief is very much less firmly established than that of a European landlord, especially when easy dismissal from office and close supervision of the treasury were such important aspects of the political system as was the case in Ashanti. Some chiefs became wealthy; some educated their children for salaried posts; but the retention of former status was much more difficult than in parallel situations in Europe. Consequently the emergent system of stratification in Africa has less to contend with in the shape of traditional authorities, anyhow in the economic sphere. The new 'class' structure is relatively open.

There are equally important implications for the economic development of the African continent. In planning the rate and progress of development we have to take into account the base line from which we start. Too many projections are less than realistic because they assume a general level of 'peasant agriculture' in Europe of 1850 and Africa of 1950.[2] This is certainly

[2] See, for example, A. F. Ewing and S. J. Patel, 'Perspectives for Industrialisation in Africa', in *Man and Africa*, ed. G. Wolstenholme and M. O'Connor (London, 1965).

not the case. In Africa the small-scale technology of Eurasia is lacking; at the village level (and 80 per cent of the population still live in rural conditions) there were wood-carvers but few carpenters, iron-workers but no mechanics, potters but no wheels or kilns. The basic craftsmen have often still to be trained.

Consequently the introduction of farming machinery is necessarily more expensive by way of overheads; inevitably, poorer use is made of tractors and similar tools. But the corollary of these technological limitations is that a great deal can be achieved with much less. Where farming has been done with the hoe, the introduction of the plough can achieve a notable leap forward. Where water has been scooped from pools and corn ground by hand, the use of small diesel pumps and grinding machines can do much to increase productivity. Where loads have been carried on the head, the wheel is a revolution.[3]

The implications of these facts are nothing new; they are known to many agricultural officers of the various countries; they have formed the basis of Schumacher's group for the promotion of intermediate (i.e. small-scale) technology. But the implications are often avoided by many modernizing politicians,

[3] See the letter from the Rev. L. Robertson, Adviser to the Ulongwe Co-operative Society, Mlanje Mission, Malawi, in *Bulletin No.* 1, 1967, Intermediate Technology Development Group Ltd., p. 13. '. . . These locally-made carts symbolise our problems in many ways. They represent a revolution—the introduction of the wheel as a means of transporting loads—but it is not seen as significant because it is not modern; it doesn't come into thinking or planning of the leaders of the economy (which is basically western). Further, the cart is too bulky per value to be worth transporting from Europe. It virtually has to be made locally. The man who sees its value is the new farmer who is part of the real revolution which is, or should be, taking place. The young leader has no experience of carts, precisely because he has missed out on the " Intermediate Stage ". '

However this general point is being appreciated by some modernizing politicians, as is shown in a recent speech made by J. K. Nyerere, of Tanzania: '. . . In many parts of the country we are beginning to follow the advice of our agricultural experts. But our major tool, the jembe, is too primitive for our present day needs. We must now abandon it and replace it with the oxen-plough. We cannot make progress by waiting until every peasant is able to possess his own tractor which he can drive and maintain. Indeed, if we wait for that we shall never leave the hoe behind us, for our present methods are too inefficient ever to produce the wealth which would enable us to buy tractors for all parts of the country, or to train the people to drive and maintain them. We are not ready for the tractor, either financially or technically; but we are ready for the oxen-plough . . .' (' After the Arusha Declaration ' speech made on 16 October 1967, Dar es Salaam, p. 4). For this reference I am indebted to Basil Davidson.

who look first to tractors rather than to ploughs, to industry rather than to agriculture, as a means of advancing their country's economy. And in this they are encouraged by the American and Soviet models, as well as by the waves of experts from East and West that flood into every corner of the continent, though rarely for long enough to appreciate the differences in the technological base, let alone the full implications of these material differences for the social system as a whole.

In placing this emphasis on differences in technology between Africa and Eurasia, differences that relate to the means of destruction as well as to the means of production, I run the risk of appearing to fall into a crude, materialistic determinism. This is not my intention; man's machinery is the product of his own inventiveness; improvements in the technology of the intellect (that is, in communications, like writing) are an important factor in the economic take-off; and changes of technology are only one factor in the process of social change. A new device, says White, merely opens the door. But that this opening up of possibilities is the most important part of the process, few can have serious doubt.[4] The plough, in particular, had a major effect in Europe where Bloch related differences in land tenure and communal obligations to the distribution of the wheelless Mediterranean plough (*aratrum*) and the wheeled Germanic variety (*charrue*).[5] In the absence of wheel, plough, and all the concomitant aspects of the 'intermediate technology', Africa was unable to match the developments in productivity and skill, stratification and specialization, that marked the agrarian societies of early medieval Europe. The so-called feudal systems of Africa lacked a feudal technology, and this absence is of critical importance in the developments of the present day.

[4] The explanation of action by means of words is necessarily subject to lineal, and often unilineal, distortion. Only more elaborate non-verbal procedures could assign a specific value to the variables involved.

[5] Bloch, 1966; White, 1940.

# REFERENCES


Ajaji, J. F. A. and Smith, R. S. 1964 *Yoruba Warfare in the Nineteenth Century*. Cambridge.

Anfray, F. 1968 ' Aspects de l'archéologie éthiopienne ', *J. African Hist.*, ix, 345–66.

Ayalon, D. 1956 *Gunpowder and Firearms in the Mamluk Kingdom; a Challenge to a Medieval Society*. London.

Barlow, F. 1961 *The Feudal Kingdom of England. 1042–1216*. (1st ed. 1955.) London.

Beattie, J. H. M. 1959 ' Checks on the Abuse of Political Power in some African States ', *Sociologus*, ix, 97–115.

1964 ' Bunyoro: an African Feudality? ', *J. African Hist.*, v, 25–35.

Binger, L. 1892 *Du Niger au Golfe de Guinée*. Paris.

Bloch, M. 1961 *Feudal Society*. (1st French ed. 1939–40.) London.

1966 *French Rural History*. (1st French ed. 1931.) London.

Bottomore, T. B. and Rubel, M. 1956 *Karl Marx: Selected Writings in Sociology and Social Philosophy*. London.

Boutruche, R. 1959 *Seigneurie et féodalité*. Paris.

Bowdich, T. E. 1819 *Mission from Cape Coast Castle to Ashantee*. London.

Bradbury, R. E. 1964 ' The Historical Uses of Comparative Ethnography. Data with Special Reference to Benin and Yoruba ', in *The Historian in Tropical Africa* (ed. J. Vansina *et al.*). London.

Braimah, J. A. and Goody, J. 1967 *Salaga: The Struggle For Power*. London.

Burton, R. F. 1864 *A Mission to Gelele, King of Dahome*. London.

Busia, K. A. 1951 *The Position of the Chief in Ashanti*. London.

Capron, J. 1965 *Anthropologie économique des populations bwa (Mali-Haute Volta)*. Ouagadougou (mimeo).

Carter, P. L. and P. J. 1964 ' Rockpaintings from Northern Ghana ', *Trans. Hist. Soc. Ghana*, vii, 1–3.

Carus-Wilson, E. M. 1954 *Medieval Merchant Venturers*. London.

Chilver, Mrs. E. M. 1960 '"Feudalism" in the Interlacustrine Kingdoms', in *East African Chiefs* (ed. A. I. Richards). London.

Cipolla, C. M. 1965 *Guns and Sails in the Early Phase of European Expansion 1400–1700*. London.

Clapperton, H. 1829 *Journal of the Second Expedition to the Interior of Africa, from the Bight of Benin to Soccatoo*. London.

Cofie, J. 1963 'The Desert of Gofan: Was it ever Densely Inhabited?', *Ghana Notes and Queries*, v, 10–15.

Cohen, R. 1966 'The Dynamics of Feudalism in Bornu', in *African History*. (ed. Butler). Boston Univ. Papers on Africa, vol. ii.

Colson, E. 1958 'The Role of Bureaucratic Norms in African Political Structures', in *Systems of Political Control and Bureaucracy in Human Societies* (ed. V. O. F. Ray), *Proc. Am. Eth. Soc.* Seattle.

Coulbourn, R. (ed.) 1956 *Feudalism in History*. Princeton.

Davidson, B. 1961 *Black Mother*. London.

Dupuis, J. 1824 *Journal of a Residence in Ashantee*. London.

Easton, D. 1959 'Political Anthropology', in *Biennial Review of Anthropology*, 1959. Stanford.

Evans-Pritchard, E. E. 1961 *Anthropology and History*. Manchester.

Ewing, A. F. and Patel, S. J. 'Perspectives for Industrialisation in Africa', in *Man and Africa* (ed. G. Wolstenholme and M. O'Connor.) London.

Fage, J. G. 1957 'Ancient Ghana: A Review of the Evidence', *Trans. Hist. Soc. Ghana*, iii, 77–98.

Fallers, L. A. 1956 *Bantu Bureaucracy*. Cambridge.

1961 'Are African cultivators to be called "Peasants"?', *Current Anthropology*, 2, 108–10.

Field, M. J. 1948 *Akim-Kotoku*. London.

Finley, M. I. 1968 ' Slavery ', *International Encyclopedia of the Social Sciences*, xiv, 307–13.

Forman, W. and Dark, P. 1960 *Benin Art*. London.

Fortes, M. 1945 *The Dynamics of Clanship among the Tallensi*. London.

Fortes, M. and Evans-Pritchard, E. E. 1940 *African Political Systems*. London.

Frazer, J. G. 1919 *Folklore in the Old Testament* (3 vols.). London.

Fustel de Coulanges, N. D. 1890 *Les Origines du système féodal; le bénéfice et le patronat pendant l'époque mérovingienne* (*Histoire des institutions politiques de l'ancienne France*). Paris.

1891 *The Origin of Property in Land* (trans. Margaret Ashley). London.

Ganshof, F. L. 1952 *Feudalism*. London. (Belgian ed. 1944.)

Gluckman, M. 1940 *Analysis of a Social Situation in Modern Zululand*. Rhodes-Livingstone Paper, 28. Manchester.

1947 ' African Land Tenure ', *Rhodes-Livingstone Institute Journal*, v, 1–12.

1960 ' The Rise of a Zulu Empire ', *Scientific American*, ccii, 157–68.

Goody, J. 1954 *The Ethnography of the Northern Territories of the Gold Coast*. Colonial Office London (mimeo).

1956 *The Social Organisation of the LoWiili*. London.

1962 *Death, Property and the Ancestors*. London.

1965 Introduction, *Ashanti and the North-west*. Inst. of African Studies, Legon (mimeo).

1966a Introduction, *Succession to High Office*. Cambridge.

1966b ' Circulating Succession among the Gonja ', in *Succession to High Office*. Cambridge.

Goody, J.—*cont.* 1967 'The Over-Kingdom of Gonja', in *West African Kingdoms in the Nineteenth Century* (ed. D. Forde and P. M. Kaberry). London.

1969 'Inheritance, Property and Marriage in Africa and Eurasia', *Sociology*, iii, 55–76.

1970 'Class and Marriage in Africa and Eurasia', *Am. J. Sociology*, December 1970.

Goody, J. and Mustapha, T. M. 1967 'The Trade Route from Salaga to Kano', *J. Nigerian Hist. Soc.*, iii, 611–16

Gravel, P. B. 1965 'Life on the Manor in Gisaka (Rwanda)', *J. African. Hist.*, vi, 323–31.

Herskovits, M. J. 1938 *Dahomey, an Ancient West African Kingdom*. New York.

Hilton, R. H. and Sawyer, P. H. 1963 'Technical determinism: the stirrup and the plough', *Past and Present*, xxiv, 90–100.

Hiskett, M. (ed.) 1963 'Abdullāh Ibn Muḥammad, *Tazyīn al-Waraqāt*. Ibadan.

Hoyt, R. S. 1961 *Feudal Institutions: Cause or Consequence of Decentralization*. New York.

Ibn Fartua, Ahmed 1926 *History of the First Twelve Years of the Reign of Mai Idris Alooma of Bornu* (trans. H. R. Palmer). Lagos.

Issawi, C. 1950 *Ibn Khaldun: An Arab Philosophy of History*. London.

Jones, G. I. 1949 'Ibo Land Tenure', *Africa*, xix 1, 44, 309–23.

Kāti, M. 1913 *Ta'rīkh al-fattāsh* (ed. and French trans., O. Houdas and M. Delafosse). Paris.

Kling, Capt. 1893 'Auszug aus den Tagebüchern des Hauptmanns Kling 1891 bis 1892', *Mitt. D. Schutzgebiet*, iv, 105–47.

Klose, H. 1899 *Togo: unter Deutscher Flagge*. Berlin.

Köhler, O. 1953–4 'Das "Pferd" in den Gur-sprachen', *Afrika und Übersee*, xxxviii, 93–110.

Kosminsky, E. A. 1956 *Studies in the Agrarian History of England in the Thirteenth Century* (ed. by R. H. Hilton, trans. by Ruth Kisch). (1st Russian ed. 1947.) Oxford.

Krause, G. A. 1928 'Haussa-Handschriften', *Mitt. Sem. Orient. Sprachen*, xxxi, 105–7.

Kyerematen, A. A. Y. 1964 *Panoply of Ghana*. London.

Legassick, M. 1966 'Firearms, Horses and Samorian Army Organisation 1870–1898', *J. African Hist.*, vii, 95–115.

Levine, D. H. 1965 *Wax and Gold*. Chicago.

Lewis, T. 1956–7 'Seebohm's Tribal System in Wales', *Econ. Hist. Rev.* (2nd ser.) x, 16–33.

Lhote, H. 1960 *The Search for the Tassili Frescoes*. London.

Loeb, E. M. 1962 *In Feudal Africa*. Bloomington, Ill.

Lombard, J. 1957 'Un système politique traditionnel de type féodal: les Bariba du Nord-Dahomey. Aperçu sur l'organisation et le pouvoir central', *Bull. IFAN*, xix, 464–506.

1960 'La vie politique dans une ancienne société de type féodal: les Bariba du Dahomey', *Cah. d'études afr.*, 5–45.

1965 *Structures de type 'féodal' en Afrique noire*. Paris.

Lyall, C. J. 1930 *Ancient Arabian Poetry*. London.

Maine, Sir. H. 1883 *Dissertations on Early Law and Custom*. London.

Mair, L. P. 1961 'Clientship in East Africa', *Cah. d'études afr.*, ii, 315–25.

1962 *Primitive Government*. London.

Maitland, F. W. 1897 *Domesday Book and Beyond*. Cambridge.

Maquet, J. J. 1954 *Système des relations sociales dans le Ruanda ancien*. Tervuren.

1961a *The Premise of Inequality in Ruanda*. London.

1961b 'Une hypothèse pour l'étude des féodalités africaines', *Cahiers d'études africaines*, ii, 292–314.

Marx, K. 1845–6 *Die Deutsche Ideologie* (trans. in *Karl Marx: Selected Writings in Sociology and Social Philosophy*, T. B. Bottomore and M. Rubel, London, 1956).

1964 *Pre-capitalist Economic Formations* (ed. E. J. Hobsbawm). London.

Mauny, R. 1952 'Essai sur l'histoire des métaux en Afrique Occidentale', *Bull. IFAN*, xiv, 545–95.

1961 *Tableau géographique de l'Ouest Africain au Moyen Age*. Mém. IFAN, No. 61. Dakar.

McNeill, W. M. 1963 *The Rise of the West*, Chicago.

1964 *Europe's Steppe Frontier*. Chicago.

Meek, C. K. 1946 *Land Law and Custom in the Colonies*. London.

Miller, J. I. 1969 *The Spice Trade of the Roman Empire*. Oxford.

Monteil, C. 1924 *Les Bambara du Ségou et du Kaarta*. Paris.

Nadel, S. F. 1942 *A Black Byzantium*. London.

Oman, C. 1924 *A History of the Art of War in the Middle Ages*. London.

Pankurst, R. 1961 *An Introduction to the Economic History of Ethiopia*. London.

Pocock, J. G. A. 1957 *The Ancient Constitution and the Feudal Law; a Study of English Historical Thought in the Seventeenth Century*. Cambridge.

Polanyi, K., Arensberg, C. M. and Pearson, H. W. (eds.) 1957 *Trade and Markets in the Early Empires*, Glencoe, Ill.

Polanyi, K. 1966 *Dahomey and the Slave Trade*. Seattle.

Pollock, F. and Maitland, F. W. 1898 *The History of English Law*, 2 vols. (2nd ed., 1952.) Cambridge.

Postan, M. 1950–1 'The Manor in the Hundred Rolls', *Econ. Hist. Rev.*, iii, 119–25.

Potekhin, I. I. 1960 'On Feudalism of the Ashanti', paper read to International Congress of Orientalists. Moscow.

Rattray, R. S. 1923 *Ashanti*. London.

1929 *Ashanti Law and Constitution*. London.

1932 *The Tribes of the Ashanti Hinterland*. Oxford.

Richards, A. I. 1960 'Social Mechanisms for the Transfer of Political Rights in Some African Tribes', *J. roy. Anthrop. Inst.*, xc, 175–90.

Richards, A. I.—*cont.* 1961 'African Kings and their Royal Relatives', *J. roy. Anthrop. Inst.*, xci, 135–50.

Rivers, W. H. R. 1912 'The Disappearance of Useful Arts', in *Festskrift Tillegnad Edvard Westermarck*, 109–30. Helsingfors.

Runciman, S. 1960 *The Families of Outremer.* London.

Schapera, I. 1956 *Government and Politics in Tribal Societies.* London.

Seebohm, F. 1883 *The English Village Community.* London.

1895 *The Tribal System in Wales.* London.

1902 *Tribal Custom in Anglo-Saxon Law.* London.

Serjeant, R. B. 1963 *The Portuguese off the South Arabian Coast.* Oxford.

Simoons, F. J. 1965 'Some questions on the economic prehistory of Ethiopia', *J. African Hist.*, vi, 1–13.

Smith, M. G. 1956 'On Segmentary Lineage Systems', *J. roy. Anthrop. Inst.*, lxxxvi, 39–80.

1960 *Government in Zazzau.* London.

Smith, W. R. 1889 *The Religion of the Semites.* Edinburgh.

Southall, A. W. 1956 *Alur Society.* Cambridge.

Steinhart, E. I. 1967 'Vassal and fief in three lacustrine kingdoms', *Cah. d'études afr.*, vii, 606–23.

Stenning, D. J. 1959 *Savannah Nomads.* London.

Stenton, F. 1961 *The First Century of English Feudalism, 1066–1166.* (1st ed. 1932.) Oxford.

Stephenson, C. 1942 *Mediaeval Feudalism.* Ithaca.

1954 *Mediaeval Institutions.* Ithaca.

Strayer, J. R. 1956 'Feudalism in Western Europe', in *Feudalism in History* (ed. R. Coulborn). Princeton.

Stubbs, W. 1882 *Chronicles of the Reigns of Edward I and Edward II*, vol. i. London.

Tomlinson, H. H. 1954 'The Language and Peoples of Gonja', MS.

Urvoy, Y. 1949 *Histoire de l'empire du Bornou.* Mém. IFAN, No. 7. Dakar.

Verlinden, C. 1955 *L'Esclavage dans l'Europe médiévale.* Bruges.

Vinogradoff, P. 1892 *Villainage in England*. London.

1911 'Comparative Jurisprudence', article in *Encyclopedia Britannica* (11th ed.), 580–7.

1920 *Outlines of Historical Jurisprudence*, i. *Introduction—Tribal Law*. London.

Weber, M. 1947 *The Theory of Social and Economic Organization* (trans. by A. R. Henderson and Talcott Parsons). Edinburgh.

White, L. 1940 'Technology and Invention in the Middle Ages', *Speculum*, xv, 141-59.

1962 *Medieval Technology and Social Change*. Oxford.

Wilks, I. 1966 'Aspects of Bureaucratization in Ashanti in the Nineteenth Century', *J. African Hist.*, vii, 215–32.

1967 'Ashanti Government', in *West African Kingdoms in the Nineteenth Century* (ed. D. Forde and P. M. Kaberry). London.

Wolf, E. R. 1951 'The Social Organisation of Mecca and the Origins of Islam', *Southwestern J. Anthrop.*, vii, 329–56.

Wrigley, C. C. 1957 'Buganda: an outline economic history', *Econ. Hist. Rev.*, x, 69–80.

# INDEX